Contents

DISCLAIMER	2
CHAPTER 1:- BUILDING MATERIALS :- WOOD AND CONCRETE	5
CHAPTER 2:- BUILDING MATERIALS :- CONCRETE VS. STEEL	24
CHAPTER 3:- BUILDING MATERIAL : GLASS	43
CHAPTER 4:- BUILDING MATERIAL :- GALVANIZED STEEL	63
CHAPTER 5:- BUILDING MATERIALS :- GALVANISED STEEL VS STAINLESS STEEL	72
CHAPTER 6:- BUILDING MATERIALS :- PRECAST CONCRETE	78
CHAPTER 7:- BUILDING MATERIALS :- ADHESIVE	97
CHAPTER 8:- BUILDING MATERIALS :- CONCRETE	113
CHAPTER 9:- BUILDING MATERIALS : PRESTRESSED CONCRETE	119
CHAPTER :- 10 BUILDING MATERIALS :- STEEL	129
CHAPTER 11:- BUILDING MATERIALS :- TMT BARS	148
CHAPTER 12:- BUILDING MATERIAL :- CEMENT	155

CHAPTER 13:- BUILDING MATERIALS :- FERROCK   161

CHAPTER 14:- BUILDING MATERIALS :- AIRCRETE   174

CHAPTER 15:- BUILDING MATERIALS FOR HUMID CLIMATE   181

CHAPTER 17:- BUILDING MATERIALS :- DRY UNDER WATER EVEN   191

CHAPTER 16:- BUILDING MATERIALS :- BLACKTOP AND ASPHALT   197

CHAPTER 17:- BUILDING MATERIAL :- CONCRETE GETS HOT   209

CHAPTER 18:- BUILDING MATERIALS :- ADMIXTURE   217

CHAPTER 19:- BUILDING MATERIALS :- STONE (MARBLE, GRANITE AND MORE..)   226

CHAPTER 20:- BUILDING MATERIALS :- M SAND   247

AUTHOR :-
GAURAV BHADANI

DIRECTOR OF BHADANI QUANTITY SURVEYING
AND TRAINING PVT LTD

https://www.constructionmanagementinstitute.com/

www.billingengineer.com

www.bhadaniinternaional.com

www.estimationandcosting.com

Copyright:

All rights reserved. No part of the site may be reproduced or copied in any form or by any means [graphic, electronic or mechanical, including photocopying, recording, taping or information retrieval systems] or reproduced on any disc, tape, perforated media or other information storage device, etc., without the explicit written permission of the editor. Breach of the condition is liable for legal action. However, the permission to reproduce this material does not extend to any material on this site, which is explicitly identified as being the copyright of a third party. Authorization to reproduce such material must be obtained from the copyright holders concerned.

DISCLAIMER

ALTHOUGH THE PUBLISHER AND THE AUTHOR HAVE MADE EVERY EFFORT TO ENSURE THAT THE INFORMATION IN THIS BOOK WAS CORRECT AT PRESS TIME AND WHILE THIS PUBLICATION IS DESIGNED TO PROVIDE ACCURATE INFORMATION IN REGARD TO THE SUBJECT MATTER COVERED, THE PUBLISHER AND THE AUTHOR ASSUME NO RESPONSIBILITY FOR ERRORS, INACCURACIES, OMISSIONS, OR ANY OTHER INCONSISTENCIES HEREIN AND HEREBY DISCLAIM ANY LIABILITY TO ANY PARTY FOR ANY LOSS, DAMAGE, OR DISRUPTION CAUSED BY ERRORS OR OMISSIONS, WHETHER SUCH ERRORS OR OMISSIONS RESULT FROM NEGLIGENCE, ACCIDENT, OR ANY OTHER CAUSE.
SEEK EXPERT ADVICE FROM PROFESSIONALS

CHAPTER 1:- BUILDING MATERIALS :- WOOD AND CONCRETE

Wood and concrete are two of the most commonly utilized building materials. These two materials have been utilized in some of the most iconic structures around the world for many years.

Wood is lighter and easier to work with than concrete, plus it is more durable and results in less thermal bridging. Concrete, on the other hand, enables for the construction of structures that are both durable and sturdy.

The remainder of this post will provide you with an overview of each construction material, which will be wood and concrete in this case. You will also learn about the advantages and disadvantages of each, as well as the elements to consider when picking building materials.

An Introduction to the Use of Wood in Construction

Throughout history, people all across the world have embraced wood construction. Wood has complicated features by nature, but humans have been able to harness these characteristics to great advantage. Boats, buildings, furniture, and interior décor are all constructed from wood.

Traditionally, wood has been divided into two categories: softwood (cone-bearing trees) and hardwood (trees that do not bear cones) (leaf-bearing). Hardwoods are commonly utilised in the building of walls, floors, and ceilings, whilst softwoods are used to make window frames, furniture, and doors. Today, engineered wood, which is commonly used in construction, has been added to the list of acceptable materials.

It takes a pretty intricate fabrication process to generate engineered wood, in which layers of veneers, wood strands, various types of wood, and fibres are linked together to form a composite material that is employed in certain construction applications. Glue-laminated timber, oriented strand board, pchapterboard, and plywood are examples of engineered woods. These materials are utilised in the construction of industrial, commercial, and residential buildings.

One of the advantages of wood that contributes to its widespread use is that it is a natural product that is both economical and readily available. Wood can be shaped and sized in a variety of ways. In addition to being renewable and providing protection against the cold, it is environmentally friendly. Aside from this, there are several pros and downsides to using wood in construction.

The Advantages of Using Wood in Construction

Wood has been utilised in construction for a long time. Despite the intention to minimise its use for environmental grounds, its advantages continue to outweigh those of competing products. The following are some of its advantages:

Toughness or tensile strength

Wood is a physically rigid and sturdy material. Furthermore, as compared to other materials, it is both flexible and lightweight. Wood has an annual ring and gain structure, which means it may be snapped or bent. However, because it is anisotropic, you will be unable to compress or extend it by pulling on the other side. Wood is quite lightweight when compared to its tensile strength.

The tensile strength of wood varies depending on the type of wood used, but in general, it allows it to hold its weight better than other materials. This decreases the need for support in various building designs and enables for the creation of larger spaces. It also makes it a great alternative for heavy building components such as structural beams because of its durability.

Insulation for electrical and thermal protection

Wood has thermal qualities that make it more resistant to high temperatures than other materials. As the temperature rises, the wood dries out and becomes more durable. It has a poor thermal conductivity, which is advantageous. This feature enables it to be used in a variety of applications across the building, including handles, doors, floors, ceilings, and walls.

The electrical resistance of wood, in contrast to other materials such as steel, makes it an excellent conductor of electricity. As a result, it is the most effective material for electrical insulation. This attribute provides a certain level of protection in households with a large number of electrical gadgets.

Sustainability

Wood is renewable in the sense that it can be grown and regrown again and again. One tree can be replaced for every one that is removed from the landscape. This enables a more sustainable use of wood that is less harmful to the environment. As a result, it is widely available locally in a large number of locations. As a result, building owners save money on transportation costs from the milling sector to the construction site.

Acoustic Characteristics

Acoustic qualities of wood include echo cancellation and sound absorption. As a result, it is widely sought after in structures where its acoustic qualities are required. Social and performance halls are examples of this type of facility. Rather than reflecting or increasing sound, wood absorbs it, lowering the level of noise in offices and living spaces, resulting in increased comfort for everyone.

The inherent beauty and visual warmth of wood are two of its most enticing characteristics. Wood has long been a favourite of architects for interior detailing, and it is now being used on exterior facades to improve the aesthetic attractiveness of buildings. There are many different kinds of wood that are utilised in construction. Softwoods such as beach, pine, ash, cedar, hickory, and birch are appropriate for use in the construction of window frames, doors, and furniture. When it comes to building flooring, walls and ceilings, hardwoods such as maple, cherry, oak, teak, walnut and mahogany are commonly employed as building materials.

Environmentally Conscious

Recently, there has been a heightened awareness of the need to minimise deforestation by reducing the amount of wood used in construction and, potentially, limiting the greenhouse effect. Wood, on the other hand, serves as a carbon storage facility, reducing the production of greenhouse gases. The policies of planting trees while you cut down others preserve the environment while also benefiting the contractor and the building's tenants.

Wood is a natural substance, and as a result, it emits lower quantities of carbon dioxide and volatile organic compounds (VOCs) (Volatile Organic Compounds). The natural organic component released by wood, on the other hand, helps to relax the occupants of a home. Biodegradable building materials include concrete and steel, which are not biodegradable. Wood, on the other hand, decomposes quickly and helps to restore the soil.

Manufacturing Is a Breeze
When compared to other materials such as concrete and steel, the production process for wood is very straightforward. This is because wood is readily available, but other materials such as concrete and steel are not. When opposed to the steel manufacturing process, the wood manufacturing method has less of an impact on the environment and produces less wastewater. Byproducts like as barks and chips can be used as biofuel in lumber mills, allowing for a reduction in the use of fossil fuels.

Wagner Meters' Moisture Management and Grade Recovery Program, for example, help lumber mills create less waste and sub-grade products while increasing their overall efficiency.

Efficiency in Energy Use

Because of the thermal insulating capabilities of wood, it is considered to be a relatively energy-efficient building material. This effectively means that it retains warmth in cold conditions, resulting in a reduction in the need for air conditioning bills. When used as flooring, it reduces the demand for heating, which is especially beneficial in extremely cold conditions during the winter months. Aside from that, because wood is easily available, the manufacturing process does not necessitate a large amount of energy.

The Disadvantages of Using Wood in Construction

Wood has long been used in building because of its natural features, but because it is a plant-based material, it is sensitive to weather and environmental conditions due to its plant-based origin. It is also subject to environmental conditions because it is a natural material.

The possibility of rot and pest infestation

Water and damp environments do not hold up well against the natural properties of wood. Even treated wood will eventually become unable to withstand moisture, making it subject to fungus, bugs, and damp rot.

When there is sufficient oxygen, fungi and pests can thrive at temperatures ranging from 25 to 30 degrees Celsius. Moisture provides them with a favourable environment in which to survive and digest it as part of their nutritional products.

Marine borers, termites, carpenter ants, and powder post beetles are just a few of the insects that cause wood deterioration as a result of drilling and driving lines. It is necessary to treat or replace wood when its functioning is affected, which can be very expensive.

Warping

Wood warps when it shrinks, swells, or bends as a result of the passage of time, variations in humidity, and temperature fluctuations. Wood, being a hygroscopic substance, will absorb vapours from its surroundings that are condensable and release moisture to the atmosphere below the fibre saturation point of the wood. When the environment alters in response to specific requirements, warping occurs, resulting in diminished functionality in areas that require precision calculations, such as window frames and doors.

Burning Possibilities

In situations where fire safety is an issue, wood is not the best choice for building materials. Wood burns quickly, and in the worst-case situation, treated wood emits poisonous compounds such as arsenic, which is lethal and can cause death when inhaled in close quarters. Despite the fact that thick wood can help to extend the burning point of the fire, engineered materials such as I-joists and oriented strand boards are extremely flammable and can spread the fire quickly.

Rapidly ageing and requiring a great deal of maintenance

Timber, if left natural and unpainted, takes on a silvery appearance as it ages and deteriorates. Wood takes a great deal of upkeep, such as treatments, repainting, and repairs, all of which are highly expensive, in order to maintain its youthful appearance and appearance. After a few years, wood becomes readily weakened as a result of environmental and weather changes, and if this is not addressed immediately, it might constitute a safety concern to those around it.

A Fundamental Overview of the Application of Concrete in Construction

Among the numerous types of buildings that concrete is used to construct are pavement and parking lots. It is also used in the construction of foundations and fences, as well as the construction of building walls, bridges, and roadways. After being mixed with water and placed, concrete begins a chemical reaction called as hydration, during which it hardens and solidifies, becoming more durable. Concrete is formed by combining cement, sand, aggregate, small stones, water, and gravel to form a stone-like material. Cement is the primary ingredient in concrete.

Historically, the Romans were the first to use hydraulic cement-based concrete, which the British later improved upon and popularised. People consume more than 6 billion tonnes of concrete every year all across the world, according to current estimates. Concrete is porous as a result of the gaps created by air voids during the mixing process and the capillary pores that form as a result of the water that is added after mixing.

By maintaining a low water-to-cement ratio, concrete is anticipated to accomplish specific properties such as wear resistance, thawing and freezing resistance, tough strength, low permeability, and watertightness. Additional admixtures are added to the concrete in order to achieve certain objectives, such as decreasing the curing time.

Concrete's Advantages in the Construction Industry

Concrete is a crucial component of the construction industry and is frequently used. When compared to other materials, concrete has a number of distinct advantages, including:

Economical

Comparing the cost of concrete manufacture to that of other engineered materials, concrete is quite inexpensive. Water, aggregates, and cement, which are the primary ingredients, are readily available in the local markets at reasonable prices. Its availability, resilience, durability, energy efficiency, and low maintenance requirements help to minimise operational and maintenance expenses, allowing it to be more cost-effective to operate and maintain. Insurance charges are also less expensive when compared to other types of materials.

Durable

Concrete develops stronger with time and lasts significantly longer than other building materials. Because concrete bonds at low temperatures, it sets, hardens, and increases strength when exposed to the elements or at typical room temperature. Concrete preserves its strength and, as a result, its long-term durability, regardless of the weather conditions. The addition of admixtures, on the other hand, can improve the strength of the material.

Energy-Efficient
Concrete has the ability to store thermal mass, which aids in the regulation of indoor temperatures as well as the reduction of cooling and heating demands by up to 8%. When combined with energy-efficient technologies like as hydronic or geothermal heating and cooling systems, as well as radiant flooring, concrete can increase energy efficiency by as much as 70%.

In the event that essential services such as water, heating fuel, or electricity are disrupted, a concrete building increases its "passive survivability" by reducing its energy demands and, as a result, enhancing the level of comfort for its inhabitants. Concrete is energy-efficient in a number of ways when it is used in pavement construction, as demonstrated below.

In comparison to asphalt pavements, concrete pavements need only one-third the primary energy for rehabilitation, maintenance, and building (according to studies). Vehicle fuel consumption and energy emissions from large trucks are reduced by up to 7% as a result of the hard surface of this material. Because of the colour of their light, there is a reduction in the heat-island effect, which results in a reduction in the need for outdoor lighting and cooling.

Resistant to both water and heat

Concrete corrosion can be caused by the chemicals present in water. The tolerance level of concrete, in contrast to the tolerance levels of wood and steel, prevents substantial deterioration and compromise in the quality of the finished product. Concrete can be used in a variety of underwater applications, including canals, pipelines, dams, waterfronts, and lining structures, as a result of this characteristic.

The corrosion of concrete in pure water is less severe than the corrosion of concrete in impure water, which contains carbon dioxide, chlorides, and sulphates as well as other impurities. Concrete is a poor heat conductor due to its porous nature. It has the ability to withstand and sustain a significant amount of heat for around 2 to 6 hours. In the event of a fire, this provides enough time for rescue crews to come and put it out.

Emissions are at their lowest.

When concrete is fully cured, it becomes completely inert and does not emit any hazardous substances, volatile organic compounds, or gases. Innovations such as Contempra, which cures concrete using carbon dioxide instead of water, result in constructions that have the lowest carbon footprint over the course of their whole life cycle. As a result, concrete emits carbon at a rate that is 6 percent less intensive than the rate of wood emission.

Versatile

Whilst solidified, concrete is useful and durable, but its fluidity allows designers to work with it to create a variety of surfaces, textures, and shapes when it is still in the mixing process. There are new developments such as photocatalytic concrete, prior concrete, and ultra-high performance concrete. These innovations have made it feasible to develop new and innovative applications while also addressing sustainability concerns.

Recycling and Adaptive Reusability are two terms that come to mind.
Concrete structures are sturdy, fire and water resistant, and they offer excellent sound absorption. They are therefore easily convertible into different types of occupancy during their service life as a result of these characteristics. The repurposing of historic structures contributes to environmental conservation and resource preservation by limiting urban development.

Concrete that has been recycled as aggregate or granular material can be used as a sub-base in parking lots, roadbeds, riprap for shorelines, and gabion walls, among other applications. The utilisation of concrete debris decreases environmental implications during new building, which would otherwise need the use of virgin materials.

Maintenance requirements are minimal to non-existent.

Painting or coating concrete structures on a regular basis is not necessary for protection or aesthetic purposes. Concrete maintains its structural integrity and shape for many years without the need for maintenance. The coatings are repainted and updated on a regular basis, resulting in a lower overall maintenance cost when compared to wood construction.

The Drawbacks of Using Concrete in Construction

Despite the fact that concrete is an extensively utilised construction material, it comes with a slew of negative characteristics. They can be adjusted by adding admixtures or by altering the concrete's ingredients and structure, but there will still be restrictions, such as the ones listed below.

Material with a semblance of bristle
As a result of its straining and softening qualities, concrete is considered to be a quasi-brittle material. Before it fails, it experiences just minor deformations that are not noticeable. Concrete has a poor toughness that contributes to its failure because to its severe low toughness. It is used in conjunction with steel to effectively increase tension and compression stresses.

Ductility and tensile strength are also low.

Because of the low tensile strength of concrete, cracks can form, and shrinkage or expansion can occur as a result of drying or moisture exposure, respectively. As a result, it is necessary to reinforce it with reinforcing bars and to incorporate structural joints to accommodate the natural expansion and contraction of the material. Because of its poor ductility, concrete is susceptible to creeping, which can result in deformation over time. As a result, while designing tall buildings that will withstand enormous loads, it is essential to think carefully about the details.

Requirement for Formwork
Formwork is required for pouring liquid concrete into a shape and sustaining the weight of the concrete. The formwork is expensive to purchase and instal, and it takes a significant amount of time and intensive labour to complete the job. Prefabrication and pre-casting are two developments that have been developed to overcome these limits.

Curing Time That Is Prolonged
Concrete must cure for a total of 28 days after it has been cast in order to achieve the requisite compressive strength. Additionally, it requires a regulated environment with an appropriate temperature for a month in order to reach its maximum potency. Curing times can be shortened by introducing admixtures or using microwave and steam curing techniques. These, on the other hand, increase the expense.

Demands highly skilled labour as well as stringent quality control

During the placement, curing, and mixing stages of concrete construction, competent workers and stringent quality control are required. This ensures that the concrete produced is of the highest possible quality. If this is not done, the concrete will have performance concerns, as well as low durability and strength. In some circumstances, specialist machinery is required, particularly during high-rise construction, in order to maintain the quality of the work and make it easier to complete.

When deciding on a building material, there are several factors to consider.
If you are building a house or a business structure, the material you choose is important not just for its overall aesthetic, but also for its strength and long-term durability. When selecting a building material, keep the following considerations in mind:

Cost-Effectiveness
When it comes to building materials, there is a big difference in price. Despite the fact that there are a plethora of materials to pick from, it is critical to conduct a cost-benefit analysis first. The cheapest material is not always the greatest material to work with. It is possible to use a cost-effective material that is within your budget as long as other requirements, such as long-term durability, are met.

Emotional appeal When choosing a building material, emotional appeal might be a critical factor to consider. The structure should be well-presented on the outside. The chosen building material should aid in the creation of the desired appearance, starting with the walls and finishings.

Types of Structures

The sort of material utilised in the construction of a structure is determined by the structure itself. A high-rise skyscraper, for example, would almost certainly be constructed of steel or concrete as opposed to other materials. Low-rise buildings are typically constructed of wood. The usage of wood in a business structure where there is a risk of fire may not be the best choice because it burns more quickly than concrete in this situation.

Availability

Generally speaking, it is preferable to use a material that is easily available in your area. The item can be delivered quickly and easily while avoiding delays in the process. Furthermore, the availability of building materials, as opposed to the necessity of transporting them over great distances, contributes to the reduction of transportation expenses.

Performance Requirements are outlined below.

It is necessary for a building material to have specific engineering features in order to work effectively. Strength, durability, soundproofing, fire resistance, and water resistance are some of the qualities of this material.

A construction material's structural qualities should be sufficient to support the weight of the building. Among its qualities should be the ability of its residents to live peacefully without experiencing any negative consequences such as chemical emissions.

Factors related to the climate and the environment

When it comes to choosing building materials, the climate is quite important. For example, due of its insulating capabilities, wood might be a good choice in extremely cold climates or during the winter months. Concrete has the ability to keep a structure cool in tropical climates or during the summer months. Overall, depending on the construction material used, this would result in a reduction in air conditioning costs during certain seasons. When it comes to construction, the environment has been ignored for decades. This is changing. But environmental issues like the utilisation of raw materials, the exhaustion of natural resources, chemical emissions, energy content, and global warming are being taken into consideration increasingly often these days, according to the EPA.

Maintenance
A building's aesthetic appeal, durability, and safety must all be maintained in order for it to remain in good condition. Materials must be chosen with consideration for their ease of maintenance, the frequency with which they will be required, and the expenditures that will be incurred....

A high-quality material will typically require less and more economical maintenance than a lesser-quality material. In the short term, low-cost construction materials may be more affordable than more expensive materials, but they will ultimately cost more in the long run.

Procedural Steps in Construction
The construction techniques for different building materials are different as well. Some may necessitate the use of specialised personnel and equipment, making them more expensive.

In other circumstances, additional work will be required on the construction site, such as cleaning debris, levelling the land, and excavating deeper foundations to ensure a more stable foundation. In addition, when a material involves extensive labour and the use of potentially hazardous machinery, the safety of workers is critical.

Supplier
When it comes to guaranteeing high-quality supplies and excellent service, having a dependable source is essential. Buildings with high-quality materials provide the intended result while also being long-lasting. A dependable supplier will also provide you with additional services such as transportation to the site and, in certain cases, delivery in the event of an emergency.

It is often advisable to work with a local supplier and to look for evaluations or recommendations from previous clients.

Sustainability
The building sector is changing at a rapid pace. While there is an increase in demand for concrete materials, there is also an increase in demand for reusable and environmentally friendly building materials. Renewable resources, such as wood, lessen the need for future creation of new materials by reducing the demand for existing ones. In addition, the construction procedure impacts whether or not the materials can be recycled.

Conclusion

Wood and concrete are two widely used construction materials that each have their own set of perks and downsides to consider. Before making a purchase, building owners should think about the following factors: maintenance, availability, supplier, climatic and environmental circumstances, the type of structure, sustainability, the construction process, and long-term durability.

Concreting offers several advantages, such as low maintenance requirements and versatility. It also has many downsides, such as a long curing period, low tensile strength, and quasi-brittleness. Wood has excellent thermal insulation capabilities, as well as being aesthetically pleasing and environmentally beneficial. It is, nevertheless, susceptible to pest infestation and rotting as a result of moisture penetration into the wood.

CHAPTER 2:- BUILDING MATERIALS :- Concrete vs. Steel

Think about how different the world would look if metal and concrete were not used in the construction of constructions. The skyscrapers that you see all over the place would not have been possible. Which of the two building materials, on the other hand, is more superior?

Concrete is widely used because it is versatile, resilient, and easy to manufacture and mould into any shape that is desired. It is also inexpensive. For its part, steel has gained popularity in the market throughout time due to its high adaptability, safety, and reliability characteristics. It is possible to utilise either one or a combination of the two materials, depending on the requirements of the project.

This chapter will assist you in better understanding the process of working with concrete and steel when creating a building structure. You will have a good understanding of the advantages and disadvantages of each product, as well as how to select the most appropriate material for each project.

Concrete: A Fundamental Overview

Commercial buildings are constructed primarily of concrete, which is one of the most frequent building materials used in their construction. It is made by mixing water and cement together, and it solidifies and hardens as a result of the hydration process. Once the mixture has been well combined with the other building components used in the construction of the bond, such as sand, cement, small stones, water, and gravel, it solidifies and sets in place.

What Are a Few of the Applications of Concrete?

Cement is a man-made substance that is used in the production of concrete. Concrete is created after cement is combined with small stones, sand, gravel, and water to form a cohesive mixture. One of its most impressive characteristics is its malleability when freshly mixed, as well as its ability to solidify and become rock-like in consistency. This feature explains why concrete may be used to construct things like pavements, foundations, motorways, overpasses, highways and bridges, and other architectural structures, as well as other types of infrastructure.

Concrete Types There are several types of concrete. There are several different types of concrete that can be utilised in the construction of concrete buildings. Some of the most commonly used varieties of concrete are as follows:

Plain Cement Concrete is a type of concrete that does not contain any additives.

Concrete made with Portland cement, aggregate, and water in a predetermined proportion is known as plain cement concrete (PCC). The most frequently encountered mixing ratio is 1:2:4. When the mixture hardens, it solidifies into a homogeneous lump that is easy to handle.

The compressive strength of structures constructed of plain cement concrete is very great; yet, unless they are reinforced with steel, the tensile strength is very low. It can be used to construct walkways, pavements, concrete walls, and other structures where tensile strength is not required.

Concrete that is light in weight

Lightweight concrete, often known as cellular concrete, is a type of concrete that flows easily. Due to the gravitational pull of the earth, the concrete levels itself. To create this type of concrete, lightweight materials such as clays, scoria, pumice, and expanded shales are combined with water to form a cohesive mixture.

The thermal conductivity of the materials is extremely low, averaging approximately 0.3 W/mK. After mixing, plain concrete develops a higher thermal conductivity, with values ranging from 10 to 12 W/mK after being exposed to air.

This form of concrete is commonly used for floor slabs, roofs, and window panels, among other applications.

Precast Concrete is a type of concrete that has been precast.
Precast concrete is cast and cured off-site, primarily in a factory setting in a controlled environment using reusable moulds. It is used in a variety of construction applications. Precast concrete has the additional advantage of being able to be manufactured to exacting specifications, which is a significant advantage.

Concrete is primarily utilised in the construction of structural components such as columns, floors, stairwells, wall panels, tunnels, and beams, among other things.

Pre-Stressed Concrete is a type of concrete that has already been stressed.

Pre-stressed concrete units are becoming increasingly common in large-scale concrete projects. In this case, the steel bars utilised inside the concrete are initially stressed before the service load is applied to them. The placement of the bars from one end of one unit to the other is required throughout the construction process.

Pre-stressed concrete combines the high-strength comprehensive qualities of concrete with the high tensile strength of steel to produce a composite material with exceptional strength. Consequently, the lower component of the structure is capable of withstanding greater tension. It is largely utilised in the construction of floor beams, roofs, bridges, heavy-duty constructions, and flyovers, among other things.

Reinforced concrete is a type of concrete that has been reinforced.
Reinforced concrete is one of the most prevalent types of concrete used in modern building and is comprised of steel reinforcement. Steel rods, wire mesh, and cables are used in conjunction with concrete to boost the overall strength of the structure. Just prior to the pouring of the concrete, steel reinforcement is installed. This sort of reinforcement, also known as rebar, is used to strengthen both the concrete's resistance to compressive stresses and the steel's resistance to tensile forces in a structure.

It has a high thermal mass and is fire resistant, which makes reinforced concrete a good choice. The material can be used to construct a wall, beam, foundation, or frame structure. Furthermore, the buildings' safety will not be in doubt, especially in the event of a fire or an earthquake.

Aside from the varieties of concrete mentioned above, there are a variety of additional types of concrete available on the market that you can employ based on the type of construction you intend to construct. They are as follows:

Concrete with a high density
Glass-reinforced concrete is a type of concrete that is reinforced with glass.
Concrete with air entrained in it
Self-compacting concrete is a type of concrete that compacts itself.
Concreting that is astute
Concrete fibre is a type of fibre that is used in the construction of concrete structures.
Polymer concrete is a type of concrete made of polymers.

The Advantages of Using Concrete as a Construction Material 1.
Concrete is one of the most frequently utilised construction materials in the world, accounting for approximately one-third of all new construction. A large part of its widespread acceptance can be attributed to the numerous advantages that it provides to any construction project. The following are some of the advantages of using concrete in a construction project.

Extremely cost-effective

When compared to other types of construction materials, the cost of cement concrete manufacture is extremely affordable. All of the ingredients are straightforward and easily accessible. To make concrete, you simply need to combine water, cement, and aggregates, all of which can be found conveniently in the market, and then pour it into a mould.

When left at room temperature, it hardens.
When exposed to regular room temperature, concrete will begin to harden almost instantly. Within a short period of time, the structure will have grown sufficient strength to maintain itself. As a result, concrete can be used at any time of day or night, regardless of the weather or time of day.

Molding and forming are simple.
The final concrete mixture has the consistency of a thick liquid. As a result, it can flow and take on varied shapes and sizes depending on the desired output. A structure's complicated shapes are best cast with this material because it is the most durable.

Highly Efficient in terms of Energy
When compared to the production of steel, concrete necessitates the use of a minimal quantity of energy. Plain cement concrete can be produced with about 450 to 750 KWh per tonne of cement required. The manufacture of structural steel consumes approximately three to ten times the amount of energy used in the production of other steel products.

It possesses exceptional water-resistance characteristics.

Concrete is more resistant to water than wood or steel, for example. Corrosion can occur, though, if the water that is being used contains chemical additives. It, on the other hand, is more resistant to wear and tear than other types of construction materials. This characteristic makes it particularly well suited for use in submerged structure development in coastal settings.

Concrete is utilised in the construction of dams, canals, pipelines, and shoreline constructions, to name a few examples. Water containing compounds such as chlorides and sulphates should be avoided at all costs in order to avoid any instances of corrosion from occurring.

The material is resistant to high temperatures. Concrete, as compared to other construction materials, can resist higher temperatures better. A component known as calcium silicate hydrate is present, and this compound has the ability to tolerate temperatures of up to 910 degrees Celsius. The material is a poor heat conductor due to its composition. So it has a limited capacity to retain heat for a particular period of time, as previously stated.

It has the ability to endure the heat of a fire for 2 to 6 hours, allowing firemen adequate time to undertake a rescue operation in the event of a fire breakout.

Allows for the use of recycled waste

A wide variety of industrial wastes can be recycled to produce normal aggregates and cement alternatives, as well as other products. Thus, concrete can assist in reducing the negative environmental effects of industrial waste when it is used. Scraps also contribute to the improvement of the structural condition of a concrete construction.

This plant does not require a great deal of maintenance. It is not necessary to paint or coat a concrete structure on a regular basis in order to keep it protected from the elements. Structures made of wood or steel, on the other hand, cannot be considered to be so. Maintenance costs are kept to a minimum because the coating only needs to be replaced when it becomes damaged.

The Disadvantages of Using Concrete as a Building Material
It possesses a low tensile strength.
Concrete has a great compression strength, yet it has a low tensile strength, despite its appearance. In engineering terms, tensile strength is defined as the amount of force required to pull something to its breaking point. Furthermore, because concrete is composed of a large number of small stones, it has cracks throughout its body. This is why steel rebar and wire mesh are added to the concrete mix to help enhance the tensile strength of the finished product.

Has a low level of toughness

Toughness is defined as a material's capacity to endure an impact. When compared to steel, it has a significantly lower ability to endure toughness than steel. Its toughness is just around 1-2 percent of that of steel, according to the manufacturer. Fiber reinforced concrete is frequently utilised in structural applications to increase the amount of toughness in a construction.

Formwork is required in this situation.
When working with concrete, formwork can aid in the moulding of various shapes based on the structural requirements. Formwork, on the other hand, can be rather expensive to purchase and instal. Furthermore, the installation can take a significant amount of time and effort. Techniques like as precasting and prefabrication are one approach to get around this problem.

It takes a long time to cure
It may only take a short period of time for the concrete to begin to dry. However, it may be necessary to cure it for up to 28 days following installation in order for it to achieve full structural strength. The use of microwave curing, steam curing, and the addition of admixtures can all help to minimise this time significantly.

Steel: A Fundamental Overview

Steel has played an essential part in construction since it was first utilised in the construction of the world's first skyscrapers in the nineteenth century. It is now playing an increasingly important role in the construction sector, as more and more cities continue to lift more and more skyscrapers to the sky. The demand for steel has increased by more than 50% in the last five years, owing to an increase in the need for buildings and infrastructure around the world.

Steel's increasing popularity can be attributed to the material's strength and longevity. It does not warp, bend, twist, or buckle in the same way as wood and concrete do. As an alternative, it is very adaptable and simple to instal. It is strong enough to withstand natural calamities such as earthquakes or hurricanes, but not as strong as concrete.

It is for this reason that steel is the most widely used structural building material in the world. It can be found in practically every type of architecture imaginable. Furthermore, it may be used for a variety of tasks and can be used in conjunction with other building materials such as glass, concrete, and galvanised flat items.

Steel for Construction: Different Types
There are many distinct sorts of steel, each dependent on the type and quantity of steel used, as well as the type of alloy. They also have a variety of physical and mechanical features that are specific to the applications for which they are used.

The chemical makeup of steel is listed below, along with the many varieties of steel.

Carbon Steel is a type of steel that has a carbon content.
Steel that contains both iron and carbon is referred to as duplex steel. Steel can be produced in a variety of grades based on the overall carbon content. When the amount of carbon in the steel is increased, the steel becomes stronger while also becoming more brittle.

Low carbon steel is less difficult to deal with than high carbon steel. For example, wrought iron steel can be used to create gates and beautiful ironwork for stairwells, as well as other applications. Structural steelwork can be constructed out of medium carbon steel. High carbon steel, on the other hand, is extremely hard and may not be bendable at all. As a result, it is most appropriate for usage in the manufacturing business.

Alloy Steel is a type of steel that is alloyed with other metals.
In order to produce alloy steel, carbon steel must be combined with one or more types of alloying elements. The goal of alloy steel production is to improve the physical qualities of the metal being produced. A steel alloy containing manganese, for example, can be used to make steel tougher and more resilient. It is also feasible to blend it with aluminium in order to provide a more homogeneous overall appearance.

The Advantages of Using Steel as a Construction Material

Steel is a popular construction material for a variety of reasons, according to design consultants and builders alike. The following are some of the advantages of employing steel in construction:

Flexibility in terms of design and strength
Steel is a powerful building material that will last for a long time, and you can count on it to do so. Additionally, it provides designers and builders with greater creative licence in terms of shape, colour, and texture. Steel, with its durability, strength, beauty, and malleability, provides architects with the opportunity to experiment with designs and come up with innovative solutions. Steel has the potential to span enormous distances, allowing for the creation of large open spaces without the need for load-bearing walls or intermediate columns.

Steel has the ability to bend to a certain radius, allowing architects to design free-form combinations for facades as well as segmented curves. In most cases, steel is a factory-finished product that has been manufactured to exacting standards in a controlled environment. The final result is predictable, and this reduces the likelihood of any onsite variability occurring during production.

It is possible to assemble it quickly and efficiently.
Steel is both efficient and resourceful, and it can be assembled quickly and conveniently at any time of year, regardless of the season. Steel requires very little on-site labour because the majority of the components are pre-fabricated. A steel structure may be erected in a matter of days by skilled workers. Building with steel can shorten construction time by 20-40 percent, depending on the size of the project being built.

Steel decreases the amount of excavation necessary since it enables for a bare minimum of contact with the earth to be established. The weight of structural steel frames is far less than that of other construction materials like as concrete, allowing them to be supported by a smaller and lighter foundation. In the building industry, the use of steel results into improved efficiency and economic value. Steel facilitates the execution of projects in a shorter amount of time.

Exceptionally adaptable
The function of a building can change in an instant. For example, a renter may wish to make changes to a building in order to create partitions or add extra rooms. It may be necessary to redecorate a building's interior and rearrange its layout in order to better serve its tenants. Steel-framed constructions can be easily transformed into a variety of various designs.

It is simple to access or make changes to existing computer networking cables, electrical wiring, and communication systems when using steel structure and flooring. It is also quite simple to convert non-composite beams into composite beams. Steel structures provide a high degree of adaptability.

Cost-Effective

Steel building allows for greater open spaces and fewer columns than other types of construction. As a result, steel is a more cost-effective option for spanning large areas. Builders have the ability to design internal rooms that are expansive and column-free. If you are building a single-story structure, steel beams can assist you in creating big open spaces. The amount of space available for column-free architecture can be extended even further in lattice or trussed structure.

It becomes easier to personalise or subdivide internal rooms when the number of columns in a building is kept to a minimum. Steel structures have a greater ability to be modified and are more versatile than other types of construction.

Fire-Resistant Materials
However, even though steel is not completely fireproof, it provides significantly superior fire protection benefits when compared to other building materials. Following extensive research to determine how structural steel responds to fire exposure, it has been demonstrated that steel provides fire resistance benefits that outweigh those provided by the majority of other construction materials. The use of steel in building allows for a reduction in the amount of fire protection that is required.

Recyclable

After a steel building has been demolished, the components of the structure can be reused or recycled. Recycling steel includes melting it down and repurposing it in a new form. Steel may be recycled an unlimited number of times without losing any of its unique qualities. Currently, recycled materials account for more than 30 percent of all new steel production. As a result, steel contributes to the reduction of the consumption of natural resources.

Adaptability in the Face of Earthquakes and Other Natural Disasters
Natural calamities such as earthquakes and hurricanes are more prone to cause damage to steel-framed structures. Earthquakes are difficult to forecast in terms of their frequency, magnitude, location, and length, to say nothing of their duration. As a result, when constructing a structure, it is critical to verify that it is capable of withstanding natural disasters. Steel is a flexible material that flexes instead of disintegrating or crushing when subjected to excessive pressure or stress.

The connections between beams and columns in a steel construction are designed to withstand gravity stresses. These connections, on the other hand, have a significant capacity to withstand lateral loads caused by earthquakes or severe winds. It is typical to discover steel buildings standing in the aftermath of an earthquake or storm, even when all of the other buildings have been demolished.

Environmentally friendly practises are becoming prevalent.

Because steel structures require less extensive foundations than concrete structures, they are lighter and have a lower environmental effect when compared to concrete structures. The steel that remains after construction can be compiled and recycled, ensuring that no waste materials are left on the job site.

Steel structures are extremely energy efficient.
Steel buildings are more energy efficient when compared to concrete structures. Because of the heat radiation emitted by steel roofs and wall panels in hot regions, using steel for construction results in a cooler indoor atmosphere in hot climes. Double steel panel walls, which are normally insulated and aid to retain heat in cold climates, can be an excellent choice for cold climates.

Aspects of using steel in construction that are disadvantageous
Despite the numerous advantages of employing steel in building, there are some disadvantages to doing so. The following are some of the drawbacks of utilising steel in construction:

Corrosion-prone materials are used in construction.

Steel is prone to corrosion due to the fact that it is an alloy of iron. Corrosion of steel reduces the functional cross-section of structural components such as columns, slabs, and beams, reducing their strength. Additionally, corrosion can have a substantial impact on the structural integrity of buildings that are constructed of steel and concrete. The steel reinforcement will begin to demonstrate an improper bond with the surrounding concrete, resulting in a reduction in the overall capacity of the unit. Anti-corrosion applications can be used to solve this problem by construction companies.

Maintenance costs are extremely high.
The maintenance costs of a steel structure may be extremely significant. Steel structures are prone to corrosion when exposed to air and water, necessitating the need for periodic painting to keep them in good condition. As a result, further painting expenses are inevitable in the future. Steel constructions will lose their thickness over time if they are not properly maintained. Steel structures may potentially lose weight by up to 35%, rendering them unable to bear external stresses and resulting in a collapse.

Susceptible to Buckling & Cracking
In addition, steel constructions are subject to buckling. Iron and steel are made up of thin plates, and the overall dimensions of steel members are lower than those of their concrete equivalents. Buckling can occur when the skinny steel parts are subjected to a great deal of pressure. Buckling is defined as a rapid collapse caused by compressive force. It is possible that using steel for columns will not be cost effective because you would need to use a significant amount to prevent buckling.

It is possible that it will not be readily available in certain areas.

Steel may not be readily available in some regions due to a lack of supply. In areas where steel is not readily available, the initial construction expenses associated with employing steel may be significantly greater than those associated with using other building materials. In many nations, higher starting prices have contributed to a fall in the use of steel in construction.

Costs of Fireproofing

Steel constructions are non-combustible, unlike wood or other materials. However, when exposed to high temperatures, such as those experienced during a fire, their strength is significantly reduced. The exposure to fire could result in substantial deformations and deflections of the primary members, which could result in the collapse of the structure. The costs of fireproofing a building might therefore be extremely high when steel is employed in the construction process.

Conclusion

Steel and concrete are both common building materials, and each has its own set of advantages and disadvantages. The choice between steel and concrete for building will be determined by a number of criteria. The considerations to consider include your budget, your preferences, and the availability of building materials, among other things.

CHAPTER 3:- Building Material : GLASS

Different Types of Glass Used in the Construction Industry

While glass does not absorb and transmit loads in the same way as steel or concrete do, it does play an important part in the natural lighting and, of course, the aesthetics of a structure.

Float glass, shatterproof glass, laminated glass, chromatic glass, and extremely clear glass are some of the most common types of glass used in the construction industry. Glass wool, glass blocks, toughened and coloured glass are some of the other types of glassware that are popular. The appropriate sort of glass for a given construction should be determined by the nature of the structure.

In this chapter, we'll go over the several varieties of glassware that are commonly used in construction, as well as their unique qualities. Ready? Then let's take a more in-depth look at each of them.

1. Glasses with a Float
To make float glass, a melt process is used in which silica sand, potash, line, soda, and recycled glass are melted together and then floated over a bed of molten tin is used. The melting of the primary materials in a furnace guarantees that the final sheet is flat and homogeneous in thickness..

The molten mass utilised in the production of float glass is allowed to harden gradually as it flows over the moulting tin during the manufacturing process. It is then annealed to reduce the various stresses that have developed as the liquid mass cools and solidifies during the process. An important benefit of annealing is that it allows glass to become more stable, allowing for a higher refractive index and density to be achieved.

Tinted float glasses are available in a variety of hues, which are determined by the colouring chemicals added during the melting process. Elements like selenium, cobalt, and iron are used to create common hues such as bronze and grey tints. Iron and cobalt are the most common metals used to create blue tints.

Float glass has a number of advantages.
Its capacity to transfer natural light from the sun makes it an excellent choice for commercial buildings because of its ability to transmit sunlight into interior spaces. Available in a variety of colours: As previously stated, the colour of float glass can be altered by adding colouring chemicals such while cobalt and iron to the mixture as it melts in the furnace. Available in a variety of colours: This increases the versatility and usability of float glass in a wide range of applications.

Float glass has a wide range of applications, and is particularly popular in the creation of architectural interiors and exteriors of buildings. Glass partitions, doors, and windows can all benefit from the application of this product. It can also be utilised as a facade in commercial buildings to improve the overall appearance of the structure.

2. Glass that is shatterproof

Because of the addition of the plastic polyvinyl butyral during the manufacturing process, this form of glass is extremely resistant to shattering. The additional element protects the glass from shattering into sharp-edged shards that could cause further breaking during an impact. Shatterproof glass is normally available in a variety of thicknesses. Lower levels can sustain significant impact, but higher levels give more protection and are capable of withstanding even greater forces. As a result, high-level shatterproof glass is widely used in skylights, floors, windows, railings, and glass staircases, among other applications.

Some of the benefits of shatterproof glass

Increased safety in the event of an emergency: Standard glass windows are susceptible to breaking when subjected to significant impacts. Since unexpected events such as flying debris and falling objects can have a significant impact on glass surfaces, it is important to plan ahead. Heavy impacts will not result in any broken or injured glass because of the use of shatterproof glass.

It does not detract from the general appearance of a building: Despite its high strength, the shatterproof glass seems to be of normal thickness, thereby enhancing the overall appearance of the structure.

Cost-effective: Unlike normal glass windows, which can shatter when subjected to significant force, shatterproof glass guarantees that structures retain their original shape even when subjected to commonplace impacts such as flying debris and high winds on a daily basis. Because of the reduction of replacement expenses, more money is saved over the long term in terms of maintenance.

3. Glass with a Laminated Surface

Laminated glass is one of the most effective types of glass that can be utilised in the construction of buildings. In order to manufacture laminated glass, two plies of normal glass are usually bonded together (strongly) with interlayers to form strong, permanent bonds. They assist in supporting the glass, allowing for greater thickness and long-term strength to be achieved.

Various thicknesses of laminated glass are available, and different glass combinations and coatings can be used to produce useful qualities such as greater insulation or low emission.

Because of the increased toughness of laminated glass, it is suited for use in applications such as glass floors, aquariums, animal cages, glass stairs, skylights, and glass ceilings. Laminated glass is especially ideal for use in buildings in high-risk zones or in constructions located in places prone to natural catastrophes such as hurricanes, such as those in Florida.

Several Benefits of Laminated Glass

Increased security: Because of the strength of laminated glass, it is nearly impossible to break through it, preventing unauthorised individuals from getting access. Because of this characteristic, laminated glass is particularly well suited for use in insecure environments.

Excellent for the environment: Laminated glass has a low emissivity, which means it reduces the amount of heat gained from the sun, hence lowering the costs connected with cooling or air conditioning systems.

Noise pollution is reduced because of the use of thick laminated glass, which is typically seen in studios and other spaces where noise must be filtered. Noise waves are disrupted by the thick layers of glass, which allows the region to be free of unwanted noise.

Because laminated glass does not shatter when subjected to high-impact forces, the likelihood of damage as a result of a breaking is considerably decreased. Increased financial security Additionally, because the glass will remain within the frame, laminated glass is safe during hurricanes and other harsh weather conditions.

In a variety of hues: Laminated glass is available in a variety of colours, shades, and tints to complement any décor. According to the preferences of the end user, it can also be constructed curved or straight.

4. Extra Clear Glass is a type of glass that is exceptionally clear.

Extra clear glass, as suggested by its name, is distinguished by its transparency and colorlessness, as well as its high clarity. This sort of glass is a one-of-a-kind float glass with an extraordinarily low iron content—hence the alternate names "low iron glass" and "ultra clear glass" for this form of glass.

As much as 92 percent of sunlight may be passed through the extra-clear glass, resulting in an exceptionally clear view. Before being melted in a furnace, silica sand with a low or no iron concentration is carefully blended with other minerals in order to produce super transparent glass. The resulting molten glass is cooled in a manner similar to that of float glass, culminating in the production of extremely transparent glass.

Extra clear glass is often utilised in interior and outdoor applications where a clear view is sought because to its enhanced clarity. It is also used in a variety of other applications. Extra clear glass is used in a variety of applications, including windows, doors, wall cladding, glass partitions, staircases, and handrails, to name a few. In addition, this form of glass is employed in solar panels because it allows for the smooth absorption of ultraviolet light.

Aside from residential applications, the purity of exceptionally clear glass makes it an excellent choice for commercial applications such as jewellery showrooms, art galleries, glass elevators, museums, and aquariums, among other things. Extra clear glass is also widely used in laminated glass, which is typically thicker than float glass because float glass is known to create an unwelcome dark green tint when exposed to sunlight.

Extra-clear glass has a number of advantages.
Extra clear glass is ideal for colour transmission since it allows for a clear view of an object without any colour distortion. This type of lighting reveals the real colour of an object, making it particularly suitable for usage in buildings located in scenic areas.
Extra clear glass has a low iron content, which reduces its reflecting qualities, allowing for the maximum amount of sunshine to pass through to the inside. This form of glass has the potential to reduce the costs associated with the use of artificial lighting.
Extra transparent glass is known to absorb heat in addition to transmitting light, making it a fantastic choice for low-temperature conditions because of its ability to absorb heat.

Flexible: Extra clear glass is one of the most adaptable varieties of glass available. Ceramic printed glass, laminated glass, tempered glass, frosted glass, insulated glass units, and tempered glass are all examples of products made from this material. Aesthetically pleasing: Homeowners who use extremely clear glass love the type of glass because of its excellent colour transmission and capacity to improve the beauty of their homes. Because of the reduced colour distortion, houses and offices can be designed in a variety of ways, based on the interests of the people who will be using them.

5. Tinted Glass is a term used to describe a type of glass that has been tinted.

Tinted glass is a particularly convenient type of glass that is distinguished by the variety of colours available. In order to manufacture tinted glass, manufacturers must incorporate color-producing elements into the glass, which aid in the addition of a little amount of colour without altering the glass's other features.

Iron oxide is known to produce a green hue when exposed to sunlight. To transparent glass, sulphur and cobalt add blue pigments, and chromium is responsible for the dark green colouring on clear glass. Uranium imparts a yellowish tint to glass, whereas titanium typically imparts a yellowish-brown colour to the same material.

Several Benefits of Tinted Glass

Tinted glass enhances total energy efficiency: Because of its energy-saving features, tinted glass is becoming increasingly popular in the construction industry. The use of a tint that is capable of absorbing heat will reduce the overall expenditures connected with heating the building.

Protection from ultraviolet rays: Tinted glass is an excellent technique to shield friends and family members from the sun's potentially dangerous ultraviolet radiation. This is due to the fact that tinted glass can absorb up to 99 percent of UV rays, ensuring that users are protected from harmful UV rays.

Improved privacy: Tinted glass is also useful when it comes to maintaining one's privacy. The beauty of tinted glass is that it can both absorb and transmit light into interior spaces while maintaining the privacy of those who reside within them. Using darker hues provides greater privacy, which is why they are chosen in commercial and residential buildings to disguise property and persons within, allowing for greater confidentiality and discreteness without the use of heavy drapes.

Maintenance-free: Tinted glass is one of the most straightforward forms of glass to keep and clean. Tinted glass, in addition to being simple to maintain, has excellent characteristics such as scratch and water resistance, which decreases the potential costs involved with frequent maintenance or replacement.

Intriguing: Tinted glass is a very appealing alternative for both commercial and residential buildings. It is not only more aesthetically pleasing to have tinted glass installed in windows and doors, but it also increases the overall value of a house by contributing to its increase in value.

Glass that has been hardened or tempered

Tempered glass, also known as hardened glass, is a form of glass that is commonly used in building because of its hardness. For the purpose of ensuring that the finished product is sturdy, producers typically treat ordinary glass with chemical or thermal treatments, which incorporate strength-enhancing qualities into the glass, so making it tougher.

Chemical and heat treatments cause tempered glass to shatter into tiny granular bits rather than shards with sharp edges, minimising the likelihood of being injured as a result of this. Tempered glass is the strongest type of glass available in terms of thermal and physical strength.

Tempered glass is particularly well suited for use in explosion-proof and high-pressure applications because of its outstanding thermal resistance, safety, and strength. This explains why it is so often used in dividers for hotels, offices, houses, and commercial structures, among other applications. Hardened glass is also an excellent choice for use in a variety of applications such as windows, doors, facades, and interior ornamental panels. Tempered glass's unique qualities also make it a popular choice for use in a variety of applications, including glass tabletops, frameless shower doors, cabinets, glass shelves, and glass near fireplaces.

Hardened/tempered glass has a number of advantages.

The ability to shape tempered glass into a variety of shapes is made possible by the adjustment of chemical and thermal properties, which can be tailored to the needs of the client. As a result, it is a fantastic choice for both residential and commercial structures, alike.

Impact resistance: When compared to float glass, tempered glass has a substantially better impact resistance, which ensures that it can endure large impacts as well as poor environmental conditions and circumstances.

Strength: Tempered glass is the material of choice for most civil engineers and architects because of its adaptability for applications that are subjected to extremely high environmental loads. While hardened glass offers a variety of advantages, it is not the ideal choice for visibility due to the tendency of the tempering process to induce optical distortion. Once toughened glass has been tempered, it becomes difficult to grind or cut the material.

7. Block of Glass

Glass blocks, also known as glass bricks, are formed when two halves that include a partial vacuum are fused together. The appearance of glass bricks varies based on the size, texture, and colour of the bricks used. The most popular glass block designs include patterned, clear, and textured faces, which make this glass type extremely versatile for use in both commercial and residential buildings alike.

Glass blocks are used in a variety of applications in the construction industry. A textured surface on glass bricks can be used in front door decorations at home because they allow for sufficient light to pass through while maintaining seclusion.

Because of their capacity to absorb light, glass blocks are an excellent choice for bathroom partitions, interior walls, and bedroom windows, among other applications. The usage of glass block as a façade is also rather prevalent. It allows sufficient light to pass through while simultaneously warming the entire structure due to its exceptional heat absorption properties.

The Benefits of Using a Glass Block

Transmits natural light: When light is transmitted through glass brick, it distorts and diffuses the rays, which helps to prevent glare. This type of glass is also available in a variety of patterns, each of which contributes to the percentage of light transmission. Glass brick, on the other hand, has good light absorption and transmission properties in general.

Excellent for privacy: When compared to conventional glass, glass brick is thick enough to absorb natural light without interfering with one's ability to remain private. In general, thicker glass blocks distort and disperse light more effectively, making it more difficult to see through them and so improving overall privacy.

Glass brick is both durable and safe, making it an excellent choice for bathroom and bedroom walls (where seclusion is essential). Its non-porous surface makes cleaning a breeze. Despite the fact that glass blocks are not load-bearing, they are nonetheless resistant to progressive deterioration and breaking, making them a safe enough long-term investment to consider in some situations.

Strong and very resistant: Because of the mortar that holds the glass blocks together and the total thickness of the blocks, glass blocks are typically stronger than normal glass. These forms of glass are extremely tough to break, making them excellent choices for earthquake-prone and insecure environments.

Excellent sound and thermal insulation: The existence of a partial vacuum aids in the retardation of heat rays, which is essential in keeping a consistent temperature inside structures. Aside from their excellent thermal insulation features, glass blocks are also well-known for their outstanding sound absorption capabilities.

8. Glass Wool

Glazed glass wool is a widely used insulating material that is created principally from molten glass as the major raw material. In addition to melted glass, other important elements of glass wool include silica sand and fixing agents, among other things. Superfine wool and loose wool are the most common types of building glass available for this use.

One of the most notable characteristics of glass wool is its adaptability to a wide range of applications. Wood frame constructions, drywall systems, steel frame buildings and cavity walls are all examples of where it can be employed. Glass wool is also extensively utilised in a variety of applications such as pipe insulation, industrial roof insulation, and sound insulation.

Glass Wool Has a Number of Advantages

Fire-resistant: Although not completely fire-resistant, glass wool can preserve its shape even at temperatures as high as 300 degrees Celsius (572 degrees Fahrenheit). Because of this characteristic, glass wool is an excellent insulator for use in both commercial and residential buildings.

Heat loss is prevented by using glass wool, which is a well-known insulator that has been used to insulate buildings for many years.

Great sound insulator: In addition to its ability to act as a thermal insulator, the qualities of glass wool allow it to be used as an acoustic barrier in barriers that are intended to inhibit sound transmission.

9. The Insulated Glass Unit (IGU).

A thermally insulated glazing unit (also known as an insulated glass unit) is recommended in locations with high energy expenses for air conditioning. This form of glass is created by creating a cavity between two or more glass panes before carefully sealing the edges together. The cavity is typically filled with a non-conducting gas such as argon or dehydrated air in order to maximise the efficiency of the insulation.

It is common practise to utilise insulated glass units in buildings that require both thermal and sound insulation. This type of glass can be used in a variety of applications, including glass roofs, double-glazed windows, skyscraper facades, highrise buildings, and skylights.

Insulated Glass Units Have a Number of Advantages

Thermal insulation is excellent because the existence of a non-conducting space between the two glass panes limits solar heat transmission from the outside to the inside of the building. During the winter months, double-glazed units are even more vital since they prevent internal heat from escaping to the structure's exterior, resulting in a more comfortable working or living environment for the occupant.

Ideal for sound insulation: Insulated glass units are also an excellent choice for soundproofing workplaces or other spaces. Because of its capacity to insulate sound, this form of glass is ideal for crowded regions, as well as for buildings located near industrial or noisy neighbourhoods.

Strong: When compared to single pane glass, insulated glass units are stronger and can sustain significant wind and snow loads, making this form of glass suitable for use in construction projects in windy and high-pressure environments.

10. Glass with a wired connection

Because of its outstanding fire-resistance capabilities, wired glass is mostly employed as a safety glass in buildings. During the manufacturing process, the glass is meticulously reinforced with robust wire mesh to increase its durability and resistance to significant impacts. The wire's function is to keep the pieces of glass in place in the event of a hit.

Because of its fire-resistant features, wired glass is often used in the windows and partitions of schools, public buildings, hotels, government offices, and other institutions. The usage of wired glass provides occupants with extra time to respond during emergencies, especially because the glass will not shatter even when exposed to extreme temperatures.

Wired glass is often used in corridors, stairwells, and emergency exits because of its fire-resistant properties. It is also utilised in the construction of skylights, roofs, and fire-resistant windows and doors. The following is important to remember: wired glass does not allow a clear vision and should not be used in situations where clear seeing is required or desired.

However, even though this form of glass can endure tremendous pressure before breaking, once it does, the sharp mesh wires that protrude from the surface can be dangerous. In addition, the wires are vulnerable to rusting, making wired glass unsuitable for use in moist or acidic conditions.

The Benefits of Using Wired Glass

Strong: Wired glass is a tough sort of glass that can take tremendous levels of pressure without shattering. Although the glass can shatter with the force of an impact, the wire mesh will hold the fragments in place, preventing burglars or criminals from entering through windows that employ this type of glass in the future.

This type of glass is ideal for use in fire crises since it is highly resistant to fire, providing residents enough time to flee in the event of a fire emergency. When it comes to fire escape routes, wired glass is an excellent choice because of its capacity to withstand fire.

Affordable: Compared to more premium solutions such as laminated glass, wired glass is a more affordable option. As a result, it is an excellent solution when looking to instal safety glass or fire-rated glass at a reasonable cost.

Glass has a number of engineering properties.

As a structural element, glass is essential in the construction of buildings because it plays an important function in transmitting natural light, insulating interior spaces, and adding to the aesthetic value of structures. Some of the most important engineering features of glass are listed here.

Transparency

In contrast to steel and concrete, glass is primarily recognised for its transparency, which allows for excellent visibility via transmitting light. Glass can be transparent from one side only or transparent from both sides, depending on the materials used in the manufacturing process to create it.

As a result, the transparency of glass is highly dependent on the materials employed, with some materials allowing for excellent light transmission while others are bad at absorbing and dispersing light. As a result, while working on projects that require natural illumination, it is preferable to use more transparent glass.

Recyclability and usability are important considerations.

Glass offers a high degree of workability when compared to the majority of other construction materials. It is the ability of glass to be melted and shaped into different shapes or types that gives it its workability. This provides architects and civil engineers with a wide range of possibilities to choose from, particularly in the design of windows and doors.

Glass is well-known for its workability, but it is also well-known for its recycling ability. In other words, glass may be used for a variety of purposes in building and can even be recycled to make concrete for use in construction projects.

Strength

Despite the fact that glass is not the strongest construction material, it can be strengthened throughout the manufacturing process to boost its overall strength. Normally, the strength of glass is determined by the value of the modulus of rupture. Considering that glass is notorious for its fragility, admixtures and laminates can be used to improve the overall strength and load resistance of glasses.

Transmittance

It is possible to define transmissivity as the fraction of visible light that goes through a piece of glass when discussing it in terms of glass. Some glasses have high transmittance and are therefore suitable for use in windows and doors, whilst others have low light absorption and distribution and are therefore unsuitable for use in windows and doors.

U-Value

It is represented by the U-value, which is the total amount of heat that can be transported through the glass. The U value of glass intended for thermal insulation should be as low as possible. When working in areas with extreme temperatures, it is preferable to use glasses that have excellent insulating capabilities. Due to the fact that the glass will either limit heat loss during frigid winters or inhibit the absorption of external heat during hot summers, it is an excellent choice.

Thermal Expansion Value is very low.

Because of its capacity to endure potentially damaging weather conditions in a wide range of climates, glass is recommended for usage in a variety of applications. Because of its excellent dimensional stability, glass is less prone than other construction materials to vary in volume as a result of temperature fluctuations. This is especially true for windows and doors.

Glass is one of the most resilient materials available for usage in both internal and external components when it comes to responding to weather variations. This is owing to the fact that glass is less prone to disintegrate over time as a result of exposure to moisture and intense sunshine.

Conclusion
As has been seen throughout the text, there are many different types of glass that are utilised in construction. There are many different types of glass to pick from, including float glass, laminated glass, extremely transparent glass, coloured glass, toughened and shatterproof glass.

The secret, though, is to find glass that is appropriate for the type of structure you are working on. When used in scenic areas where a clear view is desired, very clear glass, for example, is the ideal choice. Checking the primary qualities of each glass type is critical to ensuring that the best option is selected for usage in the construction industry.

CHAPTER 4 :- BUILDING MATERIAL :- Galvanized Steel

A wide range of applications, ranging from plumbing to interior and outdoor décor, make extensive use of galvanised steel. The corrosion-resistant and extremely long-lasting substance has a pleasing appearance. What is the manufacturing process like, and how does it differ from stainless steel?

Galvanized steel is a term that refers to steel goods that have a layer of zinc metal applied to the surface of their bodies. In order to decrease rust and make the material more resistant to all external conditions, the zinc layer is applied to the surface of the material. Galvanized steel also has the additional benefit of extending the life of projects and lowering maintenance expenses.

The remainder of this essay will take a closer look at the manufacturing process for galvanised steel. The advantages of the material will also be discussed, as well as the reasons why you should use it in your next project.

Galvanizing and Galvanized Steel: An Introduction

Galvanization is a manufacturing technique in which the goal is to coat steel or iron with a layer of zinc in order to provide it with additional protection while also reducing the likelihood of corrosion. It is possible to accomplish this using a variety of methods, but the most often used is known as hot-dip galvanising.

Galvanized steel is a term used to describe steel goods that have been galvanised, which is a zinc-coating process. Once the process is complete, you will have a product that possesses the tensile strength of steel as well as the anti-corrosion qualities of the zinc-iron coating on the surface.

A barrier between the metal beneath and the corrosive elements in the local environment is created by zinc, resulting in a higher-quality steel product with greater lifespan.

Because of the adaptability of the finished product, galvanised steel is utilised in a variety of industries ranging from construction to agriculture.

Galvanization Comes in a Variety of Forms
As we just said above, there are several different methods of galvanising steel. We'll go over them in more detail below.

Galvanization (sometimes known as hot-dip galvanization) is a process that coats steel with a zinc coating.
This is the most widely used method of galvanising steel in the world. This is what happens during the course of the process:

Steel sheets are dipped into an acidic solution to remove dirt and grime from them.
Following the cleaning process, the sheets are thoroughly rinsed with the solution.
They are then run through an acidic solution that is designed to remove iron oxides (mill scale).
Chemical cleaning agents such as ammonium chloride are used to clean the steel a second time before it is allowed to dry.
Upon drying, the sheets are immersed in molten zinc at extremely high temperatures, reaching 450 degrees Celsius (840 degrees Fahrenheit).
A link is formed between the zinc on the surface of the steel and the iron molecules on the surface of the steel.
As the sheets cool down, they become more exposed to the elements. Because of the interaction between the zinc and the oxygen in the atmosphere at this point, zinc oxides (ZnO) are formed. In addition, zinc reacts with carbon dioxide in the atmosphere (CO_2). The end product is a long-lasting zinc carbonate ($ZnCO_3$) coating that acts as a perfect barrier against corrosion.
When it comes to hot-dip galvanising, the type of coating applied to the steel can differ depending on the application. For example, galvanised steel intended for automobile parts will have a thinner covering of zinc than galvanised steel intended for construction.

Galvannealing

Galvannealing is the result of combining annealing with hot-dip galvanising to create a new product. As a result, the steel is coated with a specific coating that has a matte grey surface, which is usual. Steel galvanised in this manner is employed in construction because it is easier to weld and because it makes the material more susceptible to painting than other types of steel.

Thermal Diffusion Galvanizing is a type of galvanising in which the metal is heated to a high temperature.
The purpose of thermal diffusion galvanising is to create a zinc diffusion coating on the metals being galvanised. Known as thermal diffusion galvanising, it is a dry technique that does not require the use of any liquids. The metal to be coated is tumbled in a spinning drum containing zinc powder at temperatures of up to 300 degrees Celsius (572 degrees Fahrenheit) during which it is coated with zinc powder. Zinc diffuses into the substrate, forming a layer on top of it. These little metal goods with complex shapes are made out of galvanised steel in this sort of galvanised steel.

Electro-Galvanizing
Another way of galvanising steel is electro-galvanizing, which is becoming increasingly common. Unlike the other ways we've seen thus far, there's no need to soak the material in a hot pool of zinc before using this one. Electrical current flows into a solution of electrolyte that contains the steel in this process. This procedure is carried out at the first step of the steel-making process.

Pre-Galvanizing

Another procedure that is comparable to the hot-dip galvanising method is electrogalvanizing. However, it is performed primarily at the initial stages of the industrial process. It is necessary to clean the metal sheet before to galvanising it. Once the metal sheet has been cleaned, it is passed into a pool of molten liquid zinc and instantly recoiled to ensure that the metal is properly galvanised. The primary advantage of this process is that it assures that coils of steel are galvanised on a big scale in a timely manner, resulting in a more uniform coating than can be accomplished with traditional hot-dip galvanization.

Industry Applications for Galvanized Steel

Galvanized steel has been in use for more than a century and has been used in a wide variety of situations. This emphasises the dependability of the product. The following are some industries that make extensive use of galvanised steel.

Plumbing
Galvanized steel is used by some plumbing businesses in the construction of plumbing infrastructure. In certain cases, such systems can persist for up to ten decades, depending on the environmental conditions that prevail in the particular region. Galvanized steel is more resistant to corrosion than certain other materials and can tolerate more severe circumstances.

Automobiles and aeronautical systems

Galvanized steel is used to construct the bodies of many automobiles currently on the road. Several decades ago, galvanised metals were only used in the construction of the most luxury automobiles. The entry-barrier has been significantly lowered in recent years. Vehicle manufacturers place a high value on the material's anti-rust characteristics, which allows them to provide warranties to purchasers of their vehicles.

Galvanized materials (including steel) are employed in the manufacture of commercial aircraft parts in another industry, that is aerospace.

Agriculture
Agricultural equipment is susceptible to corrosion, as is equipment in the construction industry. Natural elements and chemicals that come into touch with these materials can swiftly degrade them if they are not protected by a protective coating of some sort. Galvanized steel indicates that farmers' equipment will be more resistant to harsh chemicals and other environmental variables when constructed of this material.

Energy from Renewable Resources
In the renewable energy market, consistency is essential because interruptions can result in considerable income losses. As a result, the producers in this area utilise materials that are well-suited for dealing with tough climatic conditions and resisting corrosion.

As a result, galvanised steel is one of the most commonly used materials in renewable energy projects. Because of the corrosion-resistant material used, the overall structure is expected to last for many years.

Construction
In the United States, the construction industry ranks among the top users of galvanised steel. Because of the material's longevity, it may be used in both residential and commercial projects during the core design stage, as well as to improve the aesthetic appeal of the finished project after it has been completed.

With its enticing lustre, galvanised steel gives any structure a contemporary and trendy look. This is why the material is widely used in fences, tubing, rails, and other applications.

Telecoms
Another area in which large companies cannot afford to utilise substandard materials is the telecommunications industry. Poor materials, in addition to generating disruptions in the entire service, can also represent a threat to the safety of personnel in the vicinity of a structure. Aside from providing the durability required for structures in the sector, galvanised steel is also a low-maintenance material, which is an additional benefit for the business.

Galvanized steel has a number of advantages.

As you can see from the information provided above, galvanised steel has numerous advantages. The following section provides a more general overview of the advantages of this material.

It has modest start-up expenses. The initial cost of galvanising steel is less than the cost of coating steel with other coating processes. Because other coating applications, such as painting, are labor-intensive, many large-scale projects opt for galvanising as their first choice, with the exception of projects with specific requirements.

It necessitates fewer maintenance efforts. This means that over the course of the life of a structure constructed using it, you will almost certainly save money. This is why, even in instances where galvanising the steel is more expensive than other forms of coating, you will receive greater value for your money over time. When you take into account the downtime caused by structures that require regular maintenance, the savings that may be realised by using galvanised steel become significant.

It has a relatively lengthy life expectancy. Across good conditions, galvanised steel has a life expectancy of more than five decades in a wide range of applications. Even in extreme conditions, such as coastal exposure, you can expect your galvanised steel to last for up to three decades with proper maintenance.

It is a dependable source of information. During the galvanization process, even the best steel companies must comply to stringent guidelines. They must adhere to a strict minimum coating thickness requirement, which ensures great performance in real-world applications. Alternatively, you can request greater standards, but galvanised steel that has been straight from the factory remains predictable and dependable.

Mechanical damage to the coating is not a problem for it. Because of the unique metallurgical structure of galvanised steel, it has a protective coating that can withstand all types of mechanical damage throughout the erection, shipping, and maintenance processes. As a result, the coating that protects it from the outdoors also means that you won't have to worry about scratches and dents to the steel during transportation. It automatically safeguards regions that have been harmed. Because of galvanization, steel will be more resistant to corrosion than other metals. As a result, when small portions of the steel are exposed as a result of damage, cathodic protection is immediately activated. This eliminates the need for touch-ups, which is a common problem with more organic steel coats.

It offers complete and total protection. All galvanised steel parts are totally protected against corrosion. Everything, from difficult locations and recesses to sharp corners, has been thoroughly covered and documented. The same level of protection will not be provided if the coating is applied to a steel structure that has already been built and completed.

Inspection is a simple process. Galvanized steel has the significant advantage of allowing you to evaluate the coating solely with your eyes. The galvanization method is designed in such a way that simply glancing at the material can tell you whether or not the coatings are sound and continuous. You can also measure the thickness of the coating in a short period of time using non-destructive and straightforward procedures.

The duration of the erection is shorter. Galvanized steel is delivered from the plant in a ready-to-use condition. This saves you time because you won't have to spend extra time on site inspecting, painting, or preparing the components for usage. In the same way, after the construction is built, you will not experience any time loss. You will be able to proceed to the next stage of construction (or begin using the structure) right away without the need for any further special procedures.

CHAPTER 5:- BUILDING MATERIALS :- GALVANISED STEEL VS STAINLESS STEEL

Due to their similar appearances, stainless steel and galvanised steel might be mistaken for one another in some situations. They are, nonetheless, distinct from one another. The next section discusses the distinctions between the two of them.

This is the Process of Preparation.
As we've seen above, galvanised steel is mostly produced by dipping steel into a tub filled with molten zinc throughout the manufacturing process. Other methods of galvanization can also be used to create it, but hot-dipping—as the process is known—is the most common and widespread.

The steel is combined with chromium in the production of stainless steels. The mixing procedure is carried out when both components are in their molten condition, and the ratio used will depend on the type of stainless steel required for a particular project. When the mixture has cooled to room temperature, it will solidify into a solid state. Acid is used to remove impurities from the surface, resulting in a clear sheen.

The Composition of the Structural Elements

Galvanized steel is composed solely of normal steel that has been coated with zinc, whereas stainless steel is composed of two independent metals that have been properly combined. As a result, the structural characteristics of both types of steel are distinct. Stainless steel, on the other hand, will contain at least 10% chromium, whereas galvanised steel is pure steel that has been coated.

The strength, pricing, and application of the product
Both types of steel are quite strong, but because of the addition of chromium to the composition, stainless steel is significantly stronger than galvanised steel. Stainless steel is significantly more expensive than galvanised steel, which is commonly used. This is why the latter is employed in large-scale projects when stainless steel would be unfeasible and would not fit within the budget constraints. Stainless steel, on the other hand, is frequently used for heavy-duty projects with a large budget.

The Process of Selecting a Galvanized Steel Company to Work With We have examined some of the most well-known galvanised steel firms above. How do you pick the best person to collaborate with? Here are some pointers to help you along the way:

Personnel Quantity and Qualitative Standards
Finding out how many workers are on the payroll of a company should be the first step in reviewing their human resource management system. The raw figures are a decent indicator of whether or not they have the staffing capacity to achieve your deadlines, but it is important to take into consideration elements such as the level of automation. A company with 50 employees and a large number of automated processes may be able to provide more quickly than a company with 100 employees and all of the work done manually.

Aside from the statistics, you must check that the employees have the talents you are seeking for in a new hire. In order to ensure that your steel parts receive unique galvanization, you must first ensure that the galvanizers can satisfy your criteria. Again, paying attention to the types of projects they have worked on in the past is the most reliable approach to determine their degree of expertise.

The Portfolio's Scope of Influence

This indicates the company's amount of experience, but it goes beyond simply determining the company's age. Does their previous experience working on steel projects in your industry qualify them to work with you? Is there a team of experts on hand that can grasp the complexities of your project's requirements? The likelihood of a brand delivering results that do not meet your expectations is significant if it does not check off these boxes.

As an example, if your company is in the automobile industry, a business with previous expertise working on architecture-related projects may not be able to deliver satisfactory outcomes to your company. Consequently, you must confirm that the organisation has both the technical ability to handle the scope of the project and the experience of working with a number of clients in your industry previously.

Pricing that is competitive
The steel fabrication process will consume a significant percentage of the budget allocated for any given construction project. You must be certain that the firm you choose to supply your galvanised steel can match your needs while also providing competitive pricing for their products. You may have quotations from a number of different companies on your desk at this point. Choosing the cheapest quote does not always imply that you are getting a competitive price.

Suppliers who use lower-quality raw materials and older equipment to manufacture your galvanised steel will almost certainly price you less than the industry standard. Alternatives that make use of higher-quality materials and cutting-edge technology may demand a higher fee, but they are more likely to fulfil your specifications and complete the project on time, saving you the time and money that would otherwise be squandered on delays and remakes.

The cost of your project will be influenced by the location of the company and the size of the organisation as a whole. As a result, when evaluating the estimates that have been submitted for your project, you must take a comprehensive approach. Deliver your basic needs to the companies and work with quotes that take those requirements into consideration. Keep an eye out for additional fees and any hidden charges.

Standards Complying with Requirements
You should concentrate your efforts on dealing with organisations that will meet the high standards you expect from them on your project. The top ones produce exceptional results on projects of all sizes, including large-scale construction. Aside from meeting the requirements of regulatory organisations, the work of the company should pass the tests specific to your organisation.

Consult with the company to ensure that they understand the manufacturing methods you desire and that they have the equipment necessary to test for compliance with your specifications.

In general, Customer Service is excellent.

You need to work with a company that has experience managing relationships with clients of all shapes and sizes. Although it may be tough to judge a company's customer service abilities from the outside, you may get a sense of how they operate from your initial conversations with them. Is it possible to find out how soon they reply to enquiries made through their designated contact channels? Do they communicate in a professional manner in written correspondence?

You can also speak with some of their previous customers to find out what they have to say about their services. A list of clientele with whom they have worked can be found on the websites of the leading companies. They should be willing to share this information with you if it is not available on their website.

The brand's resource level and financial standing are important considerations.
The most reputable steel fabrication businesses tend to be equipped with the most up-to-date metal designing and manufacturing technology. Every part of the project, from concept to final delivery to you, can be handled entirely by them because they have the means and expertise to do so. In addition, such businesses often have a large number of automated processes, which aid them in avoiding quality control problems. They will also not be afraid to use cutting-edge technology when it is required for specific tasks. You must check that the metal fabrication company you choose is financially stable—enough to spend more money where it is necessary, recruit additional professionals to enhance production capacity, and other such activities.

A financially stable company will also have a motivated workforce and will not have any dissatisfied customers or suppliers on their payroll. Ensure that the brand's suppliers are satisfied, and you can rest assured that they will employ the highest-quality raw materials possible, preventing your project from being delayed.

Conclusion
Galvanized steel is one of the most important building blocks in many industries today. The galvanization process is carried out by highly-equipped organisations that have the resources necessary to create finished steel that fulfils regulatory standards and may be customised to fit the needs of individual projects and customers.

CHAPTER 6:- BUILDING MATERIALS :- PRECAST CONCRETE

Precast concrete is a type of construction material that is created when concrete is cast in a reusable mould and then allowed to cure in a controlled environment before being delivered to the construction location. Precast concrete can be used to produce structural components such as concrete frames, walls, and floors, allowing for time and resource savings.

High quality assurance, increased time efficiency, improved construction safety, long-term durability, decreased waste, and clean working conditions are some of the advantages of employing precast concrete in construction. Fire and flooding resistance are also two advantages of precast concrete, as is its ease of dismantling and re-erection.

More information on the advantages of using precast concrete in building can be found in the following sections. Aspects of using precast concrete will be discussed as well as some of the downsides of doing so.

Precast concrete can be defined as follows:
Due to the fact that it is manufactured off-site, precast concrete earned the moniker "precast." In order to ensure that it is sturdy and capable of being used on the job site, it is typically built in numerous phases. It is possible that certain precast concrete will be prestressed with steel or cable reinforcement, depending on the project's scope, in order to boost overall structural strength.

As a result, it's important to understand if the type of structure you're developing requires additional reinforcement or not before you begin construction.

The following are some of the most important steps in the process of preparing precast concrete for use.

Engineering

Several steps must be completed prior to the placement of concrete in a mould. Detailed drawings are created by engineers using the most recent design tools and software, which are then sent to the appropriate departments for approval. After they have been accepted, the design drawings will be utilised as a blueprint for the rebar cage assembly as well as the rest of the precast construction project.

Building the Reinforcement Cage and Putting It Together
The initial stage in this process is to cut and bend steel in order to meet the specifications set in the previous phase.. The rebars must be cut with extreme precision in order to ensure that the resulting precast has the dimensions that were specified in the specifications.

Once the bars have been cut into various shapes and sizes, they are connected and carefully laced together to make the much-needed reinforcement cage for the building.

Prepare the forms and conduct a preliminary inspection before pouring the liquid.
Steel forms are used in precast factories in a variety of ways, and the type of steel form employed is usually determined by the project. A form release agent must be applied to all surfaces of the forms to guarantee that the finished product may be easily lifted from the form after the embedded items or apertures cuts have been secured.

An overhead crane scoops up the cage and inserts it in the form's core after the form has been prepared. In order to achieve high levels of precision, a professional quality control (QC) specialist must conduct a pre-pour inspection after this stage.

Placing the Concrete
Precast concrete makers must employ the highest-quality materials available in order for the project to be a success. Depending on the extent of the project, the raw materials used in concrete tend to vary, but the most typical raw materials are cement, sand, coarse aggregates, and chemical additives. They must be combined in the proper proportions to get the desired design of the finished structure. Routine raw material testing and checking on the preferred quantities for each raw material are normally carried out in laboratories in most high-quality precast companies. The spread test is one of the numerous tests that concrete must pass before it can be placed. Its goal is to ensure that the final mix has the proper flow and does not segregate during the process of mixing.

After that, the concrete is carefully lifted and set on the form, with special attention paid to ensuring that the concrete floats without entrapping air in the concrete. It is then screeded over and the appropriate finish applied after the form has been filled out.

Stripping, Curing, and Preparation
It is common for precast concrete plants to have controlled conditions, which allow the concrete to cure and attain full strength before being transported. A quality control technician will certify that the product has reached the necessary strip strength before it will be stripped (usually 2500-3000 psi).

In order to strip the precast concrete, the exterior jacket of the form is removed first, followed by the cautious collapse of the inner core to allow for the attachment of lifting devices to the concrete. Afterwards, the product is removed and thoroughly cleaned before the post-pour inspection can start.

Post-Pour
Because there is limited margin for error when working with precast concrete, it is essential to undertake a comprehensive post-pour inspection to ensure that all design dimensions are kept and that the precast concrete is free of any visible flaws or imperfections.

As soon as the precast concrete has been given a clean bill of health, plans may be arranged for its transportation to the construction site.

Not just because of its strength, but also because of its long-term durability and ease of installation, precast concrete is becoming more and more popular in modern architecture. Some of the most significant advantages related with the use of precast concrete in construction projects are listed in the following sections.

1. Highest Level of Excellence
A variety of tests must be performed in the precast plant prior to the concrete being transported to the construction site for installation. When it comes to concrete, quality assurance professionals are on the job constantly inspecting it from the time it is mixed with the other ingredients (cement, sand, aggregates) to the time it is ready to be used.

In most cases, precast concrete plants are equipped with laboratories, which aid in the analysis of the chemical contents of the raw material used. They ensure that the finished concrete is properly mixed and balanced, reducing the likelihood of an error by an order of magnitude compared to the previous batch.

Additionally, before the concrete is put into the mould, it must be inspected by quality control professionals to ensure that it has the attributes that the client has asked. Once the mould has been removed, the final product is checked for any anomalies. All of these tests and inspections help to raise the overall quality of precast concrete, guaranteeing that it meets the highest possible standards of construction..

2. Shortens the total amount of time spent building
Prepared precast concrete is often produced at precast plants or factories, where it is poured. Other tasks on the site, such as surveying and earthwork, can be completed while the construction engineers are free to work on other projects. In most cases, once the preparation of precast concrete has been completed, it is carried to the construction site, which is more often than not ready for immediate installation work to begin.

Because precast concrete can be installed while other construction operations are being carried out on the same site, it is extremely efficient. It is common for various additional jobs to have been finished by the time the precast concrete is delivered to the construction site, ensuring that the project proceeds as planned.

In addition, because precast concrete is often manufactured and cured in precast facilities, there will be no need to wait for the mixture to dry and strengthen before using it. Especially because construction begins immediately after the precast parts are brought to the site, this saves a significant amount of time for the project.

3. Using Prestressed Concrete is an option.

In order to improve the overall strength and load-bearing capacity of the final precast, precast manufacturers can use prestressed concrete in big projects with enormous loads. Large-scale construction projects benefit from the use of prestressed concrete because it helps to reduce the total size of load-bearing elements, which lowers construction costs.

Architecture and construction engineers have the flexibility and freedom to experiment with new ideas because of the ability to integrate diverse concrete types in precast plants. As a result, precast concrete is appropriate for use in a variety of construction types, including light, medium, and heavy, because it may be modified at the plant during the mixing process.

4. Safety at the construction site is improved as well.
In addition to the fact that precast concrete eliminates clutter and unneeded materials that would otherwise clog the construction site, it also has other advantages.

Furthermore, because precast concrete only requires lifting and installation once on the site, less workers will be required on the job site when compared to when concrete is cast in place on site. Because there will be fewer individuals on the job site and fewer potentially hazardous tools lying around, the lessened congestion contributes significantly to overall safety.

Because there is no need to stock raw materials on the construction site, no traditional formwork or props will be carried onto the job site. Construction sites will be safer and easier to manage as a result of this till the project is completed.

5. Extremely Long-Lasting
Precast concrete is well-known for its ability to persist for an extended period of time without requiring extensive maintenance or replacement. In most cases, high-density precast concrete is built with components that have undergone laboratory testing, which results in a product that is more durable and corrosion-resistant.

For wall, pillar, and culvert construction, precast concrete's exceptional corrosion-resistant characteristics make it an excellent choice. Due to the fact that precast concrete is water resistant, it is an excellent choice for foundation construction.

6 - Encourages the use of environmentally friendly building materials
In addition, precast concrete is produced utilising ecologically friendly techniques, making it a long-term and short-term sustainable material. For cost-cutting purposes, most precast concrete companies recycle the water that was used in the production of the concrete.

Furthermore, precast concrete is typically manufactured from natural aggregates such as sand, rock, gravel, and water, which are readily available and do not pose a threat to the environment when extracted.

Construction sites that use cast-in-place concrete frequently experience environmental damage. By utilising bracing and formwork in the manufacture of precast concrete, it is possible to reduce the amount of concrete that is needed while also reducing the trash and packing that typically accumulates throughout the casting process.

Moreover, because precast constructions often employ fewer materials, they have a lower environmental impact overall. The resulting constructions, like as walls and slabs, are typically reusable and may be readily removed from the structure if the need arises in the future.

Another feature that contributes to the long-term viability of precast concrete production is the safety of the controlled industrial conditions. Comparatively speaking, employees who work in precast concrete plants have better working conditions than their counterparts who mix concrete on the job site. This is due to the fact that precast factories can manage noise pollution, air quality, and safety hazards while also lowering costs.

7. It increases the efficiency of the project.

The fact that precast concrete is manufactured in a controlled environment contributes to its efficiency. Because precast concrete can be made in huge numbers throughout the year, there is no need to be concerned about weather conditions. Because of the ability to create precasts in large quantities (and in advance), clients can place orders and have them stored until they are needed on site. Precast concrete can also help you save money because of the fluid assembly line procedures that demand less labour than traditional concrete. The requirement to coordinate on-site labour and logistics is also decreased with precast concrete, which contributes to the overall efficiency of the construction process.

8. It improves project management.
Precast concrete provides construction engineers with the assurance that their projects will be completed regardless of the weather conditions. Weather conditions, for example, are notoriously difficult to predict, and this is especially true when project completion is concerned.

Precast concrete, on the other hand, ensures that the building team has complete control over the climate, especially because precast concrete products are typically cured in optimal conditions. Put another way, precast concrete helps to reduce the discrepancies that are caused by weather, allowing for the achievement of deadlines to be reached.

9. It is adaptable.
Depending on the design of the project, precast concrete can be shaped and sized to meet those needs. As a result, it is extremely convenient for projects that have distinctive designs.

While some may argue against the versatility of precast concrete, the key is to account for the unique shape or size of each structure during the initial stage (engineering) of the precast concrete production process. Having a clear understanding of the needed dimensions allows plant personnel to precisely cut and assemble the bars in order to create an appropriate reinforcing cage.

The shape and dimensions of the reinforcing cage must be consistent with the measurements of the moulding or form in order to ensure that the finished precast is the required size. Precast concrete is one of the most adaptable types of concrete because of its flexibility to be customised in terms of shapes, sizes, and textures.

The adaptability of precast concrete can also be shown in its capacity to be recycled or repurposed once it has been deconstructed. If the owners decide to demolish the structure, they will be able to repurpose the precast walls or structures for another purpose. Additionally, construction experts can produce everything from huge open spans to small parts with the highest degree of accuracy thanks to precast technology.

10. Appealing in terms of appearance

Given the fact that precast items are often constructed in a controlled plant setting, it is possible to blend diverse textures and colours to compliment the distinctive geometries of precast products. At every level of the manufacturing process, expert evaluation minimises the likelihood of error while simultaneously increasing the likelihood of reaching the desired result. Precast concrete is used in the construction of some of the world's most aesthetically beautiful structures, which has played a significant part in the widespread acceptance of this type of concrete. In recent years, precast concrete has grown increasingly popular for use in the construction of schools, hospitals, offices, and residential flats, among other structures.

Precast concrete has a number of disadvantages.
Even while precast concrete offers numerous advantages, it also has some drawbacks that must be considered. The following is a succinct summary of the primary drawbacks of precast concrete.

It is necessary to transport the materials to the construction site.
Precast concrete, as opposed to cast-in-place concrete, must be transported from the manufacturing facility to the construction site. The necessity to transport precast concrete results in increased transportation expenses, especially if the precast factory is located a long distance away from the job site. This might also result in variances in delivery times, particularly when dealing with terrible roads or inclement weather throughout the delivery process.

Inherently vulnerable to damage during transportation

It is common for damage to occur when precast concrete is moved from one area to another during the transportation process. Precast concrete is often heavy once it has been manufactured and cured, and it will almost surely be damaged if it is subjected to extreme impact during transit.

As a result, the production business must take extra precautions to ensure that the precast concrete arrives at its destination in good condition. The loaders and off-loaders must also exercise caution when delivering precast concrete to the job site so that big walls or slabs do not fall and cause injuries to those on the job site.

In the case of damaged precast concrete, it will take longer to complete the project, which is not ideal while working under strict deadlines.

Specialized equipment is required.
Precast concrete is naturally heavy, and lifting it requires the use of special tools such as cranes more frequently than not. When attempting to put precast concrete on high-rise buildings, a tower crane will be required for the job. Precast panels will be transported by semi-trailers to the tower cranes, where they will be gently hoisted and then positioned into their final positions.

It is possible that mobile cranes and crawler cranes will be required to lift precast concrete, which will add to the overall cost of the project. Furthermore, because there are only a limited number of cranes that can be set up on-site, the installation of the precast parts may take longer than anticipated.

The structural integrity of the building will be compromised if the installation is done incorrectly. Precast members are often put one at a time to allow them to work as a cohesive unit. Furthermore, because the seams between the precast parts have a tendency to induce structural discontinuity, poor installation may result in structurally deficient constructions.

Expert supervision is essential during the installation process since the precast concrete must be able to properly and securely transfer load pressure and forces in order for the structure to stay structurally sound. When working with precast concrete, there is limited space for error because of the requirement for meticulous attention to detail.

In order to keep precast members dry, the joints between neighbouring precast members must be sealed with particular, construction-appropriate sealants.

Precasting at a factory versus on-site
Precasting can be carried out in a casting yard, which can be located on or near the construction site in addition to the factory or plant. When there is no precasting plant close enough to the construction site, casting yards are frequently used instead of precasting plants. Furthermore, because building prices tend to rise when big items are transported across long distances, some contractors prefer to establish yards on the job site.

It goes without saying that factory precasting is preferable to on-site precasting since factories often have controlled conditions as well as laboratories for testing the combinations and ratios of the materials.

However, in order for a precasting yard to be effective, the following requirements must be met:

Material for the construction site must include raw materials such as aggregate, cement, water, admixtures, reinforcement bars, and formwork, among other things.
Formwork must be done in a separate yard from the rest of the construction.
It is necessary to have a concrete mixing area.
It is necessary to have a steel reinforcement area where rebar cages can be constructed.
It is necessary to have a casting area.
It is necessary to have a curing area.
It is necessary to have a stacking place where the final components can be stored.
Precasting concrete on site does not provide as many advantages as precasting concrete in a factory. The presence of factors such as inconsistency in the weather, which are not a concern in factory precasting, becomes a big concern when precasting concrete on the job site. When the weather is unpredictable, it can cause delays and even impair the drying of the concrete, resulting in the waste of both time and resources.

Site precasting is also known to put a strain on the construction site, resulting in clutter and other sorts of pollutants, among other things. Because of a lack of available space, the equipment required to produce rebar cages and formwork may end up posing a threat to the general safety of construction workers.

Precast concrete vs. cast in place concrete is a comparison between two types of concrete.

For many years, precast concrete has been compared against cast-in-place concrete in many applications. Although the titles suggest a significant difference, the most significant distinction is that precast concrete is produced in a controlled environment within factories. Cast in situ concrete, on the other hand, is concrete that is prepared on the construction site.

Some more distinctions between precast concrete and cast in place concrete can be found in the following sections:

Casting
When precast concrete is used, the casting of slabs, beams, and columns is usually completed in advance, allowing for significant time savings because they simply need to be set up and placed on site. Construction workers, on the other hand, must wait for cast-in-place concrete to settle, harden, and strengthen over time before continue with the project.

Inconsistencies in the casting process frequently result in delays in important activities along the critical path, making cast in situ concrete unsuitable for use in high-volume construction projects on a tight schedule.

Quality
Using the highest level of precision, precast concrete is manufactured. As previously stated in the preceding sections, precast concrete is often subjected to a number of tests by a professional quality control technician. To make matters worse, precast concrete is cured in a controlled environment within the factory, guaranteeing that the end product is exactly what was ordered.

While cast-in-place concrete can still be of the greatest quality, the success of the construction is typically determined by the level of competence of the employees on the site. When it comes to pouring concrete, skilled and experienced employees are typically preferable in order to limit the likelihood of mistakes. However, this does not ensure success because they do not have access to laboratories to test the ingredient ratios in question.

Labor
Because the majority of the work is completed off-site, precast concrete installation often necessitates only a limited number of experienced personnel. Cast in situ concrete, on the other hand, necessitates more effort in terms of preparing the components, mixing, pouring, and curing the concrete. Expert personnel are critical to the success of cast in situ concrete, which translates into higher expenditures for the company.

Construction Progress at a Rapid Pace
When precast concrete is supplied on-site, construction workers will not have to wait for the structures to gain strength because they are typically ready to be installed right after they arrive. For cast in place concrete, on the other hand, it is necessary to allow sufficient time for the concrete to settle and build strength. Cast in situ concrete typically requires a minimum of 28 days to reach full strength, which means that the building period will be lengthened as a result.

Adaptability to Environmental Conditions

Weather resistance is one of the most advantageous characteristics of precast concrete. Precast concrete structures are not prone to shrinkage, deformation, or deterioration as a result of exposure to the environment, and they can be set in place even in rainy conditions.

Concrete that has been cast in situ, on the other hand, cannot be poured when the weather is not cooperative. Furthermore, cold weather can lengthen the drying time of concrete that has been cast in place, delaying the completion of the project. When looking for work in inclement weather, construction workers must be patient and wait their turn.

Precast concrete is used in a variety of projects. During the last few decades, the use of precast concrete in modern construction has expanded dramatically.. Precast concrete is becoming increasingly popular as a result of the numerous advantages associated with its use in construction projects.

The following are some of the most notable projects that have made use of precast concrete:

Precast Concrete Structures are a type of structure made of precast concrete.
Parking spaces: Precast concrete is frequently utilised in the design of parking structures because of its durability, economy, and relative ease of installation. Concrete is used to construct a variety of structural elements, ranging from columns to pavement slabs, staircases, and even traffic barriers.

Bridges: Precast concrete materials are utilised for a variety of structural components, including girders, arches, beams, deck slaps, and caps, among others. Because of the longevity, strength, and weather resistance of precast concrete, it is an excellent choice for use in bridge construction because it requires little maintenance once it has been erected.

For culvert construction, most engineers favour precast concrete due to its resilience to water and temperature changes, which makes it perfect for wastewater and stormwater drainage applications. Because of its moisture-resistant qualities, precast concrete is also extensively utilised in the building of catch basins and curb inlets, among other things.

Buildings made of precast concrete

Precast concrete is well-known for its fire resistance, which helps to keep total fire insurance costs down. Multi-unit housing is no exception. The thick precast walls also contribute to sound absorption, making it an excellent choice for apartment buildings, hotels, complexes, and dorms, among other applications.

The use of precast concrete in the construction of schools allows for the completion of projects in a shorter amount of time, which is critical in order to avoid delays in school schedules.

Construction of retail shopping centres: Precast concrete is an excellent choice for the construction of shopping malls in both rural and suburban settings. Because of the ease with which precast concrete may be installed, investors can expedite their construction projects, provided that expert care is given in the production and installation of precast concrete.

Medical facilities: Thick precast concrete walls provide excellent sound absorption, making them an excellent choice for hospital environments. Furthermore, because of the short building time, hospitals and medical centres may be built without experiencing any unexpected delays.

Bringing It All to a Close

Precast concrete is one of the most often used types of concrete because of its high strength and long life expectancy. Another advantage of precast concrete is that it is resistant to the elements. Other advantages include ease of installation, quality assurance, and long-term durability, which distinguish it as a dependable type of concrete for both large and small scale building.

Precast concrete, despite its various advantages, is extremely prone to damage during transit, which makes it an undesirable choice. As a result, the construction crews must exercise extreme caution when unloading and erecting the precast concrete buildings on the job site.

CHAPTER 7:- BUILDING MATERIALS :- ADHESIVE

Adhesives are nonmetallic compounds that are used to permanently connect surfaces together by the use of an adhesive procedure. Construction workers can use adhesives to bind materials together and transmit load tension evenly over a junction at a lower cost than they could without them. So, what are the most popular forms of adhesives that are employed in the building industry?

Polymer adhesives, acrylic adhesives, hot melt adhesives, resin adhesives, and anaerobic adhesives are some of the most common types of adhesives used in the construction industry. Construction adhesives include electrically conductive adhesives, pressure adhesives, reactive adhesives, and plastisol adhesives, among other types of adhesives.

In this chapter, we'll look at the numerous types of adhesives that are used in construction, casting light on the advantages and disadvantages of each type, as well as how they're utilised in the field.

1. Hot Melt Adhesives are a type of adhesive that melts when heated.
In the construction industry, hot melt adhesive (HMA) is a thermoplastic glue that is frequently provided in the form of cylindrical sticks of varied diameters. The sticks are intended to be used with a hot glue gun that has a continuous-duty heating element to adhere them together.

These adhesives have a longer shelf life than other types of adhesives and can be disposed of without the requirement for specific disposal measures. Aside from that, they contain a lower concentration of organic chemicals and require no curing period.

They are utilised in the construction industry for a variety of tasks such as product assembly, profile wrapping, and laminating applications in the woodworking industry. They are also employed in the installation of electronic gadgets, particularly when it comes to the attachment of wires.

2. Adhesives made of acrylic resin

Acrylic adhesives are resin-based adhesives that are constructed of methacrylic or acrylic polymers as its primary constituents. Acrylic adhesives are incredibly strong and effective when it comes to forming numerous bindings. As a result of its resilience to environmental variables like as intense sunlight and moisture, it is widely used in the building industry.

There aren't many adhesives that can compete with acrylic adhesives in terms of strength and capacity to hold goods in place. This is owing to the high cohesion and adhesion capabilities of acrylic adhesives, which ensure that things are held firmly in place during the application process. Acrylic adhesives are available in two different forms: paste and liquid. In most cases, liquid acrylic adhesives are applied with a brush or a moist cloth, and they are most commonly employed in carpentry or upholstery applications, as well as in difficult-to-reach locations during the construction process. Paste acrylic adhesives are extensively utilised in everyday building projects and are well-known for their ability to form strong bonding surfaces.

3. Epoxy Resin Adhesives (also known as epoxy resin adhesives)

Dimensional stability, great mechanical strength, chemical resistance, and affordability are all characteristics of epoxy resin adhesives that have long been recognised. Steels, non-ferrous metals, aluminium, fiber-reinforced composites, ceramics, bricks, foamed constructions, glasses, and woods are all excellent candidates for bonding with epoxy adhesives because of their strength and durability.

Epoxy adhesive systems are composed of a basic resin, an accelerator, a hardener, fillers, flexibilizers, additives, and diluents, among other components. The chemical resistance and thermal stability of epoxy resin adhesives are determined by the base resin that is used in their manufacturing.

Epoxy adhesives are a wonderful choice for both light and heavy construction work because of its longevity, strong tensile and shear strength, and low shrinkage. However, although the bindings formed by epoxies are typically hard, they can be made more flexible by modifying the components used in the manufacturing process.

When epoxy adhesives are used to join two dissimilar metals together, the bond line functions as a barrier, preventing galvanic corrosion from occurring between the metals. Epoxies are extremely resistant to moisture, fuel, and chemicals, making them an excellent choice for use in the construction industry. Because of their excellent electrical insulating qualities, epoxy adhesives are also employed in the electrical industry.

Whether used in exterior construction or interior decoration, epoxy resin adhesives are available in either one-part or two-part epoxy systems, which ensures that extra firm connections are formed when used in exterior construction or interior decoration.

Anaerobic Adhesives are the fourth type of adhesive. Anaerobic adhesives are made up of dimethacrylate monomers that cure only when there is no oxygen present. They are used in the construction industry. Anaerobic adhesives, as compared to other types of adhesives, are less poisonous, gentle, and non-corrosive to metals, characteristics that make them particularly well suited for use in the construction industry.

Anaerobic adhesives, in addition to dimethacrylate monomers, also contain multifunctional and monofunctional ester monomers, stabilisers, a redox radical initiator system, several types of modifiers, and resins.

Some of the most significant advantages of employing anaerobic adhesives are their low odour levels, durability, and ability to cure on demand. Anaerobic adhesives, on the other hand, are not recommended for use on porous surfaces or thermoplastics, despite their capacity to form strong and long-lasting connections.

5. Adhesives Under Pressure
When it comes to bonding to a surface, pressure adhesives rely significantly on viscosity and flexibility. The name suggests that pressure adhesives function when pressure is put between the adhesive and the point of attachment, and this is exactly what happens.

This form of glue is often used on stickers, however it cannot be relied upon to keep two different substances together in the same way. In the construction industry, it might be useful for adhering cautionary notices such as danger signs or on-site instructions to walls.

6. Adhesives that are electrically conductive
In electronic applications, electrically conductive adhesives are frequently employed, particularly in situations where electrical current must be transferred between two or more components. This type of adhesive cures in less than two minutes and is composed primarily of conductive pchapters, which account for 80 percent of its composition.

It is most often a 2-component epoxy, however polyester and acrylate are also typical choices for the base adhesive. The cost of the conductive components used in electric conductive adhesives is often determined by the total number of components employed. Iron is commonly used in low-cost electrically conductive adhesives since it is a poor conductor of electricity. Copper or silver, on the other hand, is used in more expensive ones.

7. Plastisol Adhesives are the seventh type of adhesive.

The polyvinyl chloride (PVC) pchapters of plastisol adhesives are scattered in a plasticizer, which makes them stick together. They offer excellent peel resistance as well as high flexibility, which make them particularly well suited for use in the construction industry. Plastisol adhesives, on the other hand, are particularly sensitive to shear stress, which is their primary drawback. Plastisol adhesives are likewise prone to creeping when subjected to extremely large loads, as previously stated.

Plastisol, despite being extremely effective in bonding, is not ecologically friendly, which explains why alternatives such as epoxy resins are becoming increasingly popular. Plastisol adhesives are also useful for bonding non-pretreated metals together because of their ability to absorb oil.

8. Reactive Adhesives are the eighth type of adhesive. Reactive adhesives, also known as curable or chemically hardening adhesives, are adhesives that require a chemical reaction to attach two surfaces together. There are two types of reactive adhesives: one-component and two-component reactive adhesives. Because of their capacity to form strong connections, reactive adhesives are particularly well suited for use with large loads that require high strength adhesion and permanency.

8. Reactive adhesives, in addition to curing quickly, have the capacity to tolerate extreme climatic conditions, making them excellent for use in outdoor structural applications.

Reactive adhesives, despite the fact that they are extremely durable and powerful, must be applied precisely and with the appropriate amount of precision in order to form strong bindings. As soon as they are exposed to air, reactive adhesives begin to cure immediately, necessitating the utilisation of gear or dispensers when utilised in building projects.

9. Solvent-Based Adhesives The performance and effectiveness of solvent-based adhesives are determined by the polymer system that is utilised in their formulation. Being that most solvent-based adhesives include flammable solvents, it is essential to exercise caution when using them to avoid the possibility of a fire.

Immediately after the application of a solvent-based adhesive, the solvent begins to evaporate (very quickly), increasing the total viscosity of the adhesive layer. While adhesives can be used to form bonds, it is preferable to allow the solvent to evaporate before applying the glue to ensure strong bonding.

Thermoset Adhesives are number ten on the list. Thermoset adhesives are plastics or polymers that are utilised as adhesives because they are thermosetting. This sort of glue is typically given uncured and is composed primarily of monomers that have not been bonded together. Treatment with curing agents like as light, heat, or a chemical hardener may be required in order to cure.

In most thermoset adhesive systems, the resin and hardener are sold separately and then combined to begin the curing process in a mixing tank. Healing time ranges from ten minutes to more than an hour depending on the situation.

Temperature- and moisture-resistant Thermoset adhesives are well-known for their high strength, excellent heat and moisture resistance, and excellent gap-filling capacity. One-component adhesive systems are also utilised to connect surfaces together, despite the fact that two-component adhesives are extremely common.

Epoxy is a fantastic example of a thermoset adhesive because of its superior gap-filling and moisture resistance capabilities. Epoxy is often used to join wood because of its superior gap-filling and moisture resistance properties. Its ability to form strong bindings on glass and metal makes epoxy resin a popular choice for usage in the building industry.

The following are some examples of thermoset adhesives:

In addition to being known as phenol formaldehyde resin (PF), this glue is also known as phenolic resins and is often used in laminating materials such as wood and paper. It is typically found in plywood that is intended for outdoor use.
Polyimide adhesives are used in a variety of applications. With the ability to withstand exceptionally high temperatures of up to 500°C (932°F), this adhesive is particularly well suited for use in the construction of members with excellent electric and thermal conduction qualities.

Polyester resin (also known as polyester resin): In the bonding of composite constructions, particularly fibreglass, this type of glue is frequently employed as a less expensive alternative to epoxy resin. When compared to epoxy, polyester resin is less heat and moisture resistant, and it forms a slightly weaker bond than epoxy.

As a result of its high bonding strength and long-term stability, thermoset adhesives are among the most widely used adhesives in the construction industry. They're frequently utilised for heavy applications on building sites, and they may necessitate the employment of heavy machinery during the application process.

Adhesives that are based on water

Water-based adhesives, often known as waterborne adhesives, are manufactured from either soluble synthetic polymers or natural polymers, depending on their application. Various synthetic polymers, such as cellulose ethers, polyvinyl alcohol (PVA), carboxymethylcellulose, methylcellulose, and polyvinylpyrrolidone, are used to make soluble synthetic polymers.

Natural polymers can be obtained from a variety of sources, including plants (starches and dextrins), animals (bones and hides), and sources of protein (fish, blood, casein, milk albumen, and soybean).

Water-based adhesives can be supplied as dry powders or as solutions, depending on the formulation. When working with dry powders, it is necessary to mix them with water before applying them. This is due to the fact that water-based adhesives gain their bonding power when water is lost through absorption or evaporation during the bonding process.

Although most water-based adhesives have a lengthy shelf life, it is important to remember that extended storage is not possible with most of these products.

Water-based adhesives, despite being extremely inexpensive, flexible, and strong, require a little extra attention in order to optimise overall performance. These adhesives may have a strong odour and may even react when they come into contact with the skin of the user. It does, however, dry fast and may be applied in a roll or spray form, making it a simple way to glue diverse surfaces together during construction.

Adhesives that cure in the ultraviolet light

UV curing adhesives, also known as light curing adhesives, are adhesives that cure by utilising light and natural radiation to initiate the curing process. UV curing adhesives form lasting connections without the use of external heat sources because of the free radical chemistry involved in the process. UV curing adhesives are available in a variety of viscosity ranges and chemical system configurations (mostly polymer-based). These adhesives have the ability to bond a wide range of substrates, even those that are diametrically opposed to one another, without reducing overall strength.

UV curing adhesives are among the most widely used materials in the construction sector because of their versatility and low cost. In addition to glass and plastic, it may be used to bond cabinet handles and shower doors together.

Adhesives based on Phenolic Resins
In most cases, phenolic resins are manufactured in the form of films or liquid compositions, and they are made by condensing formaldehyde and phenol. Phenolic resins are often used to glue objects together because of their ability to penetrate and adhere to a wide range of fillers and reinforcements, as well as their low cost.

In addition, phenolic resin adhesives are extremely compatible with cellulose fillers, making them suitable for bonding wood products such as plywood, pchapterboard, oriented strand board (OSB), and hardboard together. When it comes to creating strong bindings that are both chemical and thermal resistant, this type of glue excels.

One significant benefit of phenolic resin adhesives that makes them particularly well suited for use in construction is their ability to retain tight connections even when exposed to adverse weather conditions. Although phenolic resin adhesives are among the most expensive on the market, their strong bonding and tenacity are well worth the price of admission.

Reduced localised stress is one of the benefits of adhesive bonding.

A common problem with most assembly procedures is that the load and stress distribution is unequal. However, by adding adhesives to the surfaces that require bonding, the load is equally distributed over the surfaces, resulting in a reduction in localised stresses. When compared to alternative assembly procedures, adhesives also perform exceptionally well in terms of fatigue resistance.

Mechanical Shock Resistance that is unmatched
When exposed to harsh environments for an extended period of time, screws and nails tend to deteriorate. Adhesives, on the other hand, can be designed to withstand harsh environmental conditions because of their specific chemical features, hence extending their useful life.

The ability to bond materials that are similar and dissimilar.
While certain assembly processes can create strong bindings, there aren't many that can compete with adhesives when it comes to attaching things that are distinct in nature. Adhesives can be used to join metals and plastics, glass and wood, and even ceramics together. Adhesives are a wonderful choice for building projects because of their unique bonding qualities. This is especially true when it comes to joining surfaces composed of diverse materials.

Disassembly is a breeze.

The ease with which adhesive bonding can be removed when necessary makes it superior to other assembly techniques such as mechanical fastening. Removing nails and screws from a building's structural integrity can be dangerous, especially when extra strain is applied to places that are particularly vulnerable to failure.

There are numerous removal processes available for adhesive bonding, which can help to guarantee that the bonded surfaces are properly separated without causing any damage to the bonded surfaces.

Contours
When opposed to mechanical fasteners, adhesive-bonded surfaces tend to be more visually pleasing in appearance. This is due to the fact that, unlike rivets, adhesives do not leave any exterior projections or gaps when used. Injury from the exterior projections is particularly dangerous for joints that are exposed.

Durability
It is commonly known that adhesives have long-term durability and dimensional stability. Due to the large number of products available on the market, it is simple to select the one that is best suited for each type of construction project. Adhesives for minor construction work can be found on the market, while others are better suited for bonding surfaces that are intended to distribute severe loads and strains.

Versatility

As can be seen from the examples of adhesives discussed above, there are numerous types of adhesives that are classified according to the components that are utilised in the construction process. Because a variety of items are available, contractors have the freedom and flexibility to select the products that are most appropriate for the nature of their projects.

Even better, technological improvements and intensive study have assisted in the development of new recipes for adhesives that are exceptionally strong and long-lasting. As a result, adhesives are one of the most adaptable assembly materials to have on a building site, as some have excellent electrical conductivity properties while others have amazing thermal resistance characteristics.

Improve the strength of adhesive bonds by following these steps.
When working with adhesives, it is vital to ensure that the surface is properly prepared. The importance of surface preparation cannot be emphasised, especially given the fact that surfaces are typically polluted with grease, dirt, dampness, oil, and other impurities. As a result, it is critical to clean the surfaces before to construction in order to limit the likelihood of failure throughout the process.

Metals such as copper and aluminium generate oxide coatings on their surfaces that cling tenaciously to their substrates, resulting in surfaces that are ideal for the application of adhesives. Surface treatment is frequently required for glass in order to create and maintain strong enough connections. Failure to properly prepare surfaces prior to applying adhesives can result in bond failure, which can be quite costly when working on highly complicated joints.

As a result, it is critical to carefully review product specifications to ensure that the proper sort of glue is utilised. There are several reasons for this. For example, certain applications demand for electroconductive adhesives, while others benefit from water-based adhesives or UV curing adhesives. As a general rule, the optimum type of adhesive should be determined by the nature of the building and the sort of bond that is needed.

CONCLUSION

As has been demonstrated in the preceding discussion, there is a diverse variety of adhesives, each of which has its own set of characteristics. Epoxy resins, for example, are frequently utilised in the construction industry because of their weather resistance, durability, and temperature resistance, among other characteristics.

Other types of adhesives are also suitable for use in building projects, with some, such as electrically conductive adhesives, allowing for the passage of electric current via the joint between the two materials. As a result, while working with adhesives, the key is to select products with features that are suited for creating strong bindings that will endure the test of time.

CHAPTER 8:- BUILDING MATERIALS :- CONCRETE

Concrete is one of the most important building materials, and it is often used in pavements, foundations, roads, bridges, walls, fences, and poles, among other applications. Concrete, after being mixed with water, solidifies and hardens, aiding in the bonding (strongly) of other construction elements. With so many different varieties of concrete to pick from, it is critical to understand which one is most appropriate for the building project you are working on.

Plain concrete, high-density concrete, lightweight concrete, precast concrete, reinforced concrete, prestressed concrete, shotcrete, air-entrained concrete, self-consolidating concrete, and glass-reinforced concrete are the most common types of concrete used in construction. Plain concrete, high-density concrete, lightweight concrete, precast concrete, reinforced concrete, prestressed concrete, shotcrete, air-entrained concrete, self-consolidating concrete, and glass-reinforced concrete

Continue reading to learn everything you need to know about the many varieties of concrete that are used in building. Ready? After that, let's get right down to business.

What Exactly Is Concrete?
Concrete is an important construction material that is utilised in a wide range of construction projects, whether they are small, medium, or large. Concrete, as a composite material, is made up mostly of water, cement, and coarse aggregate (sand, gravel, or rock).

When the ingredients listed above are combined in the proper proportions, they usually form a stone-like paste that hardens over time, keeping the various components of a construction together.

Fresh concrete can be used in a variety of applications since it can be easily formed into rectangles, squares, circles, and a variety of other forms, depending on the sort of project being undertaken. It is impossible to exaggerate the importance of mixing cement, water, and aggregates in the proper proportions. This is due to the fact that the ratio of components used in the production of concrete dictates the strength, workability, resistance to the elements, and durability of the finished product.

Generally speaking, cement and lime are employed as binding ingredients, with sand serving as the fine aggregate in the mix design. Crushed stones, clinkers, gravel, and broken bricks are used to finish the composite by functioning as coarse aggregates, and they are used to complete the composite.

Similarly, as technology improvements continue to have an impact on construction techniques and procedures, so does the evolution of concrete. Concrete is often classified based on three factors: the types of materials used in its construction, the stress conditions under which it was constructed, and the overall density. The following section contains an in-depth examination of the many varieties of concrete that are often used in building.

1 PLAIN OR ORDINARY CONCRETE

Plain concrete is one of the most regularly used types of concrete and is an excellent choice for pavements and other applications where tensile strength is not required to be exceptionally high. In order to combine cement, sand, and aggregates in this concrete type, the 1:2:4 mix scheme is used.

Despite the fact that plain concrete has a decent durability rating, it is not known to hold up well to wind loading and vibrations, which is why it is only used for light to medium building projects.

2. Constructed of lightweight materials

Lightweight concrete is defined as concrete with a density of less than 1920 kg/m3 and is used in a variety of applications. This form of concrete is noted for having a low thermal conductivity, which makes it ideal for protecting steel structures that are exposed to the elements.

Lightweight concrete is extremely flowable (because to its low density) and self-leveling, which means it does not require a great deal of effort to be applied and spread evenly across the surface of the concrete slab. Because of its lightweight nature, it is suitable for use in the construction of floor slabs, roofing, and window panels, among other things. In lightweight concrete, the most typically used aggregates include scoria, pumice, perlite, vermiculite, clays, and expanded shales, among other materials.

3. High-Density Concrete is another option.

High-density concrete, often known as heavyweight concrete, has the highest density of any of the concrete varieties available. When mixed and manipulated throughout the construction process, this type of concrete has a density of 3000-4000 Kg/m3, making it difficult to work with. Construction engineers must employ crushed rocks as the coarse pchapters in high-density concrete in order to achieve the desired strength and durability. The most widely used coarse aggregate material in concrete is barytes, which has a specific gravity of 4.5 and is a strong performer.

Generally speaking, high-density concrete is utilised in huge constructions such as atomic power plants, which generate potentially dangerous radiations, due to its density and strength. As a result of the high-density concrete composition, radiations will be prevented from flowing through the walls. This sort of concrete is usually expensive, and it requires excellent mixing in order to be strong enough to support its own weight.

4.Reinforced concrete is the fourth type of concrete.

This form of concrete, also known as reinforced cement concrete (RCC), is made up of different-sized steel bars that operate as reinforcement to boost the overall tensile strength of the structure. Therefore, with reinforced concrete, wires, cables, and steel rods are typically installed before the concrete has a chance to harden.

As a result of enhancing tensile resistance, steel reinforcements make concrete stronger and more resistant to compressive forces than it would otherwise be. When it comes to reinforced concrete, professionals are usually concerned with ensuring that the link between the reinforcement and the settling concrete is as strong as possible. A successful bond between the concrete and the steel reinforcements allows the resulting reinforced cement concrete to withstand a wide range of stresses and loads in small, medium, and large buildings alike. This places RCC among the most significant concrete kinds, especially given the fact that it may be employed in a wide range of projects without compromising the structural integrity of the structure.

5. Precast Concrete

Precast concrete, as implied by the name, is typically prepared and cast off-site, typically in a controlled manufacturing environment, rather than on site. Precast concrete is frequently utilised in projects where it is necessary for the concrete to have well-balanced ratios in order to hold other elements in place.

Moreover, in order to maximise the effectiveness of the concrete during construction, precast concrete is typically manufactured in factories, with careful attention paid to the proportioning of the key constituents of the concrete mix. By the time concrete is delivered to a job site, it has typically been expertly mixed to meet the exact specifications of the building project being carried out.

It is popular to employ precast concrete for precast walls, beams, tunnels, columns, staircase modules, poles, and concrete blocks, among other things. The primary advantage of employing this form of concrete is the speed with which it can be assembled on-site. Additionally, because the materials are prepared in controlled circumstances, the finished concrete is typically of the greatest possible quality, according to the manufacturer.

CHAPTER 9:- BUILDING MATERIALS : PRESTRESSED CONCRETE

As a result of its ability to handle huge loads and tensions, prestressed concrete is frequently employed in large-scale building applications. This type of concrete allows for the insertion of specified engineering stresses on the primary components, which can be used to assist offset the potential stresses that are expected to occur once the load is applied to the structure.

As a result, prestressed concrete is effective because it combines the strong compressive strength of concrete with the high tensile strength of steel. Construction engineers may be assured that the concrete will keep its weight and, as a result, will effectively neutralise the forces generated during the construction process if the concrete is prestressed before usage.

Prestressed concrete is extensively utilised in a variety of heavy construction applications, including piles, floor beams, water tanks, bridges, runways, and railway sleepers, among others.

The following are some of the advantages of using prestressed concrete:

The compressive strength of the produced concrete is increased.
In order to reduce the likelihood of tension cracks in the lower sections of load-bearing beams, prestressed concrete is essential.
Prestressed concrete also contributes to increased total shear resistance, which reduces the requirement for stirrups.
It is particularly well suited for projects with substantial dead loads, such as long-span roofs, bridges, and other large-scale buildings.
Air-Entrained Concrete is the seventh type of concrete. Concrete with microscopic air bubbles, known as air-entrained concrete, is a special type of concrete that differs in size based on the manner of preparation. The air bubbles serve the purpose of forming expansion chambers, which allow water to expand as it freezes, so reducing the internal pressure of the concrete.

Concrete specialists use foaming chemicals such as alcohols, fatty acids, and resins during the mixing phase of the concrete to entice the air into the mix. However, in order for the job to be effective, engineering supervision is required in order to verify that the air-entrained concrete is properly blended.

In areas prone to severe snowfall and subsequent freeze-thaw cycles, air-entrained concrete is a frequent construction material. Air-entrained concrete is therefore more resistant to deterioration caused by freezing and thawing cycles, scaling, and abrasion than traditional concrete constructions.

8. Concrete that hardens in a short period of time
Rapid hardening concrete, also known as rapid-set concrete, is an excellent choice for projects that must be completed in a short amount of time. This type of concrete has a well-deserved reputation for setting quickly and being extremely resistant to low temperatures, making it excellent for usage at any time of year, regardless of the weather conditions.

Because of the greatly shortened hardening time, the concrete can bond various members in a short period of time, allowing projects to be completed on schedule.

Even though it is one-of-a-kind due to its traits and characteristics, this form of concrete is best suited for specific applications such as cold weather and tiny repairs that must be completed quickly. The use of rapid hardening concrete as the primary concrete in large construction projects is not recommended because it will almost likely affect the overall structural integrity of the building or bridge.

9. Constructed of Glass

With the use of recycled glass as the aggregate in glass concrete, it is possible for the finished construction to have a sparkling, elegant aspect that adds to its overall aesthetic appeal. In addition to increasing the aesthetics of a project, glass concrete is recognised to provide superior thermal insulation when compared to other varieties of concrete.

This form of concrete is most commonly utilised in construction projects where the final appearance is of particular importance to the client. The majority of the time, glass concrete will be seen on decorative facades and large-format slabs. The reflective or coloured appearance of the inlaid glass lends a reflective and colourful appearance to the slab or exterior cladding.

10. Asphalt Concrete

This form of concrete, which is also known as blacktop or asphalt, is typically used on roadways, highways, walkways, parking lots, and airport runways, among other places. Asphalt concrete is created by combining asphalt and aggregates in a way that allows for rapid hardening without sacrificing strength or durability.

Because of its skid resistance, workability, durability, stability, permeability, and flexibility, asphalt concrete is an excellent choice for usage in regions subjected to heavy dynamic loads. Asphalt concrete, on the other hand, must be properly blended in order to endure automotive and human traffic.

11. Shotcrete

To prepare shotcrete concrete, use the same steps as you would for regular concrete. The fundamental distinction between shotcrete and other types of concrete, on the other hand, is the method by which it is applied.

Shotcrete is typically applied utilising nozzles because they allow for the release of high pressure. Using nozzles to set up the concrete means that the concrete will begin compacting soon after it is sprayed on the surface, which is advantageous.

This form of concrete is best suited for tiny repairs to wood, steel, or concrete structures, but it should not be utilised as the principal concrete for a structural foundation. Shotcrete concrete is especially advantageous when working in tough-to-reach regions since the nozzle allows construction workers to shoot the concrete at challenging angles that would otherwise be impossible.

12. Pumped Concrete

Pumped concrete is commonly utilised when working on high-rise buildings, particularly on the upper floors that are difficult to reach. This type of concrete is often very workable and can be easily transported to the upper levels using a pipe system that is installed in the basement. In many cases, the pipe is a rigid or flexible hose that is used to carefully discharge the mixed concrete to the desired locations on the difficult-to-reach floor.

Pumped concrete is not only used in high-rise buildings; it is also used in a variety of other applications. Concrete can also be used in small-scale construction projects such as swimming pools, as well as large-scale construction projects such as bridges and motorways. Additionally, when it comes to creating exceedingly level floors on horizontal building projects, pumped concrete can be of great assistance.

However, only fine pchapters are utilised in order for the pumped concrete to flow up the pipe and into the designated regions. It is possible to ensure that concrete flows smoothly to its assigned places by using fine aggregates.

13. Limecrete

Limecrete is a type of concrete in which lime is used in place of cement, as the name suggests. Limecrete is a lightweight concrete that is commonly used to construct domes, vaults, and floors. It is made up of lightweight materials such as sharp sand and glass fibre and is primarily utilised in the construction of domes, vaults, and floors.

Cleanliness and great recyclability are two characteristics that distinguish this type of concrete as environmentally friendly. Limecrete can also be used to make floors that have radiant floor heating installed beneath them.

14. Fiber-Reinforced Concrete (FRC)

A fiber-reinforced concrete mix is used to increase flexibility, tensile strength, and durability, among other beneficial properties, in order to provide excellent load-bearing capabilities. The fibres are often manufactured from a variety of materials, including glass, steel, polymer, carbon, and even coconut fibre in some cases.

Furthermore, because some fibres have been shown to react with cement, it is critical to exercise caution during combining. Fiber-reinforced concrete is primarily employed in industrial floors, airport runways, and bridge pavements, among other applications. When fibre is added to concrete, the concrete's resistance to cracking is increased, making it perfect for high-traffic areas.

There are several advantages to using concrete. Concrete, as one of the most regularly used building materials, has a wide range of applications that vary based on the project type and the environment. Some of the most significant advantages of using concrete in construction are listed below.

Concrete is a very cost-effective material.
The basic components of concrete, such as water, cement, and aggregates, are usually readily available and reasonably priced. In general, when compared to other construction materials such as steel and polymers, concrete is less difficult to come across.

Concrete can be formed into a variety of shapes and sizes.

Fresh concrete has a great flowability due to the fact that it is still in a liquid form (depending on the mixing ratio). Depending on the complexity of the project, the resulting concrete can be poured into a variety of different configurations called formworks. Concrete, on the other hand, is one of the most convenient building materials to work with because of its capacity to be shaped and sized in a variety of ways.

Construction Concrete exhibits excellent water resistance characteristics.
Despite the fact that dissolved compounds in water such as chlorides and sulphates can cause progressive corrosion in different concrete types, properly mixed concrete outperforms steel and wood when it comes to corrosion resistance. This explains why concrete is often utilised in submerged applications like as dams, canals, pipelines, linings, and shoreline constructions, as well as in the construction of retaining walls.

Concrete is capable of withstanding high temperatures. Concrete offers superior heat-resistance capabilities when compared to both wood and steel construction. In part, this is because the primary binder in concrete, calcium silicate hydrate (C-S-H), can tolerate temperatures of up to 910 degrees Celsius (1670 degrees Fahrenheit).

Concreting is a poor conductor of heat, which permits it to store a substantial quantity of heat from its near surroundings. Apart from that, concrete can endure high temperatures for up to 6 hours, giving rescuers adequate time to react in the event of an emergency fire. Because of concrete's exceptional fire resistant qualities, it is frequently used to fireproof steel in construction.

Concrete Has a Positive Impact on the Environment
Did you know that concrete can recycle a wide range of industrial wastes and utilise them as alternatives for aggregate and cement in the production of new concrete? In addition to fly ash and waste glass, slag and even ground vehicle tyres are some of the elements that are widely employed.

The ability of concrete to reuse waste material contributes to environmental conservation and the promotion of green construction.

Application with a number of different modes
Concrete distinguishes itself from other construction materials by virtue of the large range of application strategies available. A construction worker can apply concrete to the desired locations by hand, pouring it, spraying it, grouting it, or pumping it.

Furthermore, concrete varieties such as shotcrete dry quickly and solidify quickly, making concrete a versatile material for a wide range of construction tasks, including highway construction.

Many Different Types of Concrete Are Available
As has been demonstrated throughout the essay, concrete is available in a variety of forms. The availability of a variety of concrete types ensures that the structural integrity of the building is not affected during the construction process. This is due to the fact that there is concrete for little repairs, concrete for large projects, and even concrete for light tasks.

Concrete is a low-maintenance material.

The care required for concrete constructions is minimal in comparison to other building materials, which require frequent painting and coating to defend against the effects of the weather. Maintaining concrete's strength and durability over a period of several years with regular coating renewal makes it an excellent choice for large building projects.

CONCLUSION

Concrete is a highly convenient construction material due to the fact that it is available in a variety of shapes and sizes. There are many different varieties of concrete to choose from, including high-density concrete, light concrete, precast concrete, prestressed concrete, and even glass concrete.

One of the most effective types of concrete is one that is complementary to the design of the project currently under development. In order to ensure that a particular concrete type's properties are considered prior to its use in building, it is necessary to do so. Heavy construction projects will necessitate the use of concrete that is both strong and durable, with high tensile and compression resistance qualities.

CHAPTER :- 10 BUILDING MATERIALS :- STEEL

Was it ever brought to your attention that steel accounts for 50% of global demand in the construction industry? Steel is used extensively in the construction industry, and as the world's population continues to grow, there will be an increasing demand for infrastructure and new buildings. What are the advantages of employing steel in construction projects?

Steel has a number of advantages in the building industry, including its adaptability, durability, availability, and low cost of production. It is also 100 percent recyclable, which makes it an environmentally responsible alternative for a variety of construction projects of varying complexity.

Continue reading to learn more about the advantages of using steel in building. We will also go over some of the areas in which steel is used in construction and how it is used in buildings in greater detail.

1. Availability of resources

Steel has become easier to come by as a result of overproduction, making it more readily available than most other metals. It is available in two forms: pre-fabricated components and raw alloy. It is possible to purchase fabricated parts such as frames and beams that have been manufactured by suppliers both locally and globally. If you don't have the competence to create these parts, you can concentrate on the building aspect of the project and order steel parts whenever it is convenient for you.

The best aspect is that, once you have a construction plan in hand, you can order steel components. This will assist you in double-checking the measurements and determining the optimal storage solution. Due to the fact that much of the steelwork is prefabricated, the construction process is quite short.

2. It is light in weight.
Despite the fact that steel is a durable construction material, it is one of the most lightweight. That is why a steel structure is relatively light when compared to a concrete structure of equal size. Steel's rigidity and tremendous strength are credited with this achievement.

Steel also has the advantage of being lightweight, which makes it easy to carry and hoist with a crane. Because of its lightweight nature, it can be used to reconstruct structures. When the attributes of weight and versatility are combined, it becomes an excellent building material for anyone who may require extension in the future. As a result of the high strength-to-weight ratio, the foundation and other structural support systems of your construction will be easier to design. However, it is important to note that, despite the fact that steel is lightweight, it must be securely fastened to the foundations in order to prevent the material from being blown away by the wind.

3. The ability to adapt.
For those looking for a construction material that allows them to be more creative, steel is a good choice. You have complete control over the material and can make any changes you see fit at any moment. Steel frame allows for the removal of walls with relative ease, which makes renovations more straightforward.

Architects are given the freedom to come up with innovative solutions and to experiment with new concepts. Steel can be used to produce free-form combinations as well as segmented curves and shapes.

Because steel beams are so strong, you won't need any internal supporting walls once you get them installed. This means that the building can be utilised for whichever purpose you desire. Such structures are ideal for a variety of functions, including shops, offices, warehouses, garages, barns, and religious institutions.

Because the area inside is open, your imagination and creativity are the only things that can constrain you. The nicest aspect is that you may split a large room by adding inside walls to make it feel more manageable. It's simple to change things when you decide to pull down and shift the walls around in your home or office. Because there are no load-bearing walls, the technique is straightforward.

Furthermore, the availability of ready-made structural components such as C-sections, I-sections, and angle sections contributes to the diversity of the design.

4. Affordability is important.
Steel construction will save you money in the long run. This is due to the fact that steel is easily available. In addition, you will save money due to the reduction in building time and personnel costs. The steel pieces and frames are made offsite with precision, which reduces the amount of trash produced by the process.

When employing lumber, you'll need to purchase more cubic volume than usual. This indicates that you will have approximately 40% wastage during the manufacturing process. When you use steel, you will require fewer personnel, resulting in lower labour costs. The best thing is that steel is a ferrous metal, which means that you may sell it as scrap metal if you so want.

Steel is long-lasting, which means it requires little to no upkeep over time. Furthermore, it has the ability to withstand seismic activity, high winds, fire, and other natural catastrophes. Materials such as plastic and wood have a limited useful life span and can be extremely expensive to replace when they do. Taking all of these considerations into consideration, it becomes clear that steel is a reasonably priced construction material.

However, you should keep in mind that a steel structure is not always the most cost-effective option. The functionality and kind of construction have a significant impact on the cost of the project. When compared to the amount of money you would spend on concrete or wood, steel construction of a tiny building will be more expensive. When constructing a large span structure, the usage of steel is highly recommended.

5. The ability to withstand adverse weather conditions
Steel structures are resistant to dampness and the elements. Nonetheless, depending on the carbon percentage of the steel material you pick for construction, this element may be compromised or even eliminated. Rust resistance can be improved with the use of a powder treatment and hot zinc coating. This will ensure that the structural steel component will be resistant to the effects of moisture or water in the future.

6. The ability to cover a large amount of ground quickly.

Steel construction gives you a competitive advantage because the material can span greater lengths, such as with steel ceiling joists. Engineers and contractors can produce a large amount of space with steel parts, as opposed to other types of materials.

Especially useful for working on large-scale projects such as industrial structures. For long open spans, steel is a good choice because it is lightweight and durable. When working with wood or concrete as a structural support, this would be impossible.

A rectangular shape is characteristic of prefabricated steel buildings, which are both cost-effective and durable. When you employ steel, on the other hand, you can create an internal space with a clear span and greater floor-plan flexibility. You will have more flexibility for future renovations as a result of this.

7. Construction Time that is Concise
Another advantage of employing steel in building is that you can save time on the project. When it comes to building, time is of the essence, especially if you are working on a project that is time-sensitive. You don't want to go over budget or run over the deadlines that have been set.

Steel frames might help you save time in the construction process. In order to finish a large-scale project on short notice, all you need is pre-engineered structural designs from steel manufacturers. Handling and installation are made easier by the use of steel parts. When compared to typical building erection time, pre-engineered steel buildings require one-third less time for installation. Furthermore, quality control is performed throughout the fabrication process. So you won't have to be concerned with the possibility of human error, which will save you time by reducing down on the amount of time spent measuring, cutting, and installing. Savings on site preparation, lower interest rates, and a faster return on investment are all possible as a result of the compressed construction schedule.

This type of savings can amount to between 3 and 4 percent of the total project cost. As a result, cash flow will be enhanced, and working capital requirements will be lowered.

Durability and sturdiness are the eighth and last criteria. Despite its lightweight nature, steel is one of the strongest and most durable metal available. In contrast to a construction built of traditional wood, a steel construction will be lasting and robust. To top it all off, steel can survive a wide range of adverse weather conditions, including high winds and earthquakes, as well as heavy snow and other snowfall, among others.

The rust-resistant feature of steel contributes to the overall endurance of the material. You won't have to be concerned about pests like bugs and termites causing damage to your structure. Buildings constructed of steel are more fire resistant than those constructed of wood.

Furthermore, you will not have to worry with the rot caused by moist wood. In this case, you'll have to replace the rotted wood. Steel, on the other hand, is not vulnerable to rotting, which is a significant advantage.

Ductility is the ninth characteristic.
When a material is ductile, it means that it can sustain extreme deformation without cracking or breaking. Steel is a material that retains its shape even when subjected to tensile pressures. When it comes to assembling the various components of your building, this is a distinct advantage. Materials with low ductility will crack when subjected to the pressure of the welding junction while being heated.

Steel has a ductility quality that allows it to deform without collapsing and to bend out of its original shape without losing its strength. The fact that steel structures are rarely destroyed by earthquakes explains why they are less vulnerable to damage.

10. Fire Resistance

Steel buildings and structures are more fire resistant than wooden frames when compared to other materials. Steel construction minimises the likelihood of a fire occurring in any building. There is a particular flame-retardant coating that can be applied to structural steel to boost its endurance.

11. Energy Consumption that has been optimised

Your energy expenditures might be reduced by up to 50% if you have a steel building that is insulated. This is due to the fact that metal buildings require a variety of equipment to keep the temperature stable during different seasons.

12. The ease with which it can be maintained

The most advantageous aspect of using steel is that the cost savings continue after the construction project is completed. This is due to the fact that steel has a longer lifecycle and is therefore less expensive to maintain. The fact that it is long-lasting allows you to save money as well. Adding paint to a surface also means that you will not have to deal with rust and corrosion as much.

13. Expansion and service integration

It is also possible to generate additional space while employing steel in construction, which is another advantage. Mechanical ventilation can be installed without interfering with the depth of the floors or the original layout of the building. In situations when there are height constraints, this is extremely useful. Furthermore, because of the adaptability of steel, the structure's lifespan is significantly increased.

14. Environmental Friendliness is important.

The steel recycling organisation steel.org estimates that more than 80 million tonnes of steel are recycled each year in North America. The research goes on to indicate that recycling steel saves enough energy to power almost 18 million households every year, according to the findings.

Steel is the only material that can be recycled several times without losing its strength. It is also the most expensive. Steel recycling is connected with lower carbon emissions and a reduction in overall energy consumption. The use of steel in construction helps to conserve natural resources. Furthermore, when you use steel, everything is fabricated off-site, which reduces labour costs. That implies you'll be dealing with a clean construction site and a reduced carbon footprint during the construction process.

As another example, if an iron or steel structure is demolished, the components can either be recycled or returned to the steelmaking process to be used in the creation of new features. Because steel buildings are lighter than concrete structures, they do not necessitate the use of substantial foundations as do concrete structures. You can either retrieve steel pile foundations and reuse them at the end of a structure's lifecycle or recycle them if you plan to employ steel pile foundations. Modern steel buildings also incorporate solar energy, such as photovoltaics, which contributes to the overall environmental friendliness of the structure. Wood and concrete, on the other hand, are almost completely non-recyclable.

Temporary structures will appreciate the convenience of this feature.

When constructing a steel structure, you can utilise a variety of procedures such as welding or riveting. These structures are simple to deconstruct and do not provide many difficulties. As a result, these types of designs are useful when you don't have a lot of time to devote to construction.

Pest Resistance is number sixteen.
Termites are attracted to wood and other natural materials. Despite their small size, these little pests may wreak havoc and inflict structural damage, putting the entire structure at risk. Bugs, on the other hand, are unable to infest steel structures.

17. Overall, the construction quality has been improved.
Buildings constructed of concrete or wood have a tendency to lose their lustre over time. Steel structures have a long service life and require little maintenance since the material is dimensionally stable and is not impacted by weather fluctuations or other natural disasters—steel structures provide a long service life while requiring little maintenance and saving you money.

18.Social Benefits
Steel is used in construction to help support a local workforce, which ultimately translates into supporting family life and establishing a stable neighbourhood. Because of the short construction timescale, any negative influence of steel-based constructions on the surrounding community is minimised. Furthermore, the construction process and the use of limited site delivery have the potential to benefit the entire community.

Steel in Construction Has a Long and Proud History

When railroads were a well-known means of transportation, steel emerged as a prominent construction material for the first time. Cast iron, wrought iron, and steel were the types of metals that were often used in the 1800s. Wrought iron was popular among blacksmiths and those working in the construction industry, while cast iron was used in farming and cooking, and steel was utilised for high-end products such as swords and watches. It wasn't until 1855 that Sir Henry Bessemer came up with the Bessemer process, which was designed to increase the efficiency of steel production. Companies were able to produce steel with high tensile strength because to their efforts. Despite this, wrought iron remained the preferred material for iron-based construction.

Sidney Thoma devised a method of removing phosphorous from steel in 1879, which was patentable. Steel quality improved as a result, and as a result, steel prices could not be made much more reasonable. The volume of output increased over time. By the 1880s, the consistency of steel quality had been established.

The Great Chicago Fire, which occurred in the United States, destroyed timber-framed buildings. As a result, people in charge were forced to enact stringent building codes that mandated the use of non-combustible building materials such as marble, stone, and brick. Other building solutions, such as wrought iron and steel structure, could be used by contractors as well.

The Home Insurance Building in Chicago was the first structure to be constructed with a steel skeleton frame and concrete walls reinforced with steel rebar. It was too light to withstand the elements and had to be dismantled in 1931.

A more durable and sturdy steel product was developed in the early 1900s as a result of technological developments. By the 1920s, steel output had increased to 60 million tonnes per year, establishing the United States as the world's leading producer of steel. Steel was increasingly being used in the construction of bridges, railroads, office buildings, and factories. With the passage of time, steel became the material of choice for the majority of construction projects.

As steel building technology advances, the applications for steel are becoming increasingly diversified. Steel is currently seen in a variety of places, including private garages, medical institutions, and skyscrapers. Steel is now divided into two types of applications: straight wall applications and arched applications for pre-engineering with additional benefits.

Steels that are commonly used in construction
The following are the most frequent types of steel used in construction:

Stainless steel is a type of steel that is corrosion resistant.
Carbon steel is a type of steel that has a carbon content.
Steel rebar is used in construction.
Steel for structural purposes
Stainless steel with alloying elements

Stainless Steel is a type of steel that is corrosion resistant.
The corrosion- and strain-resistant qualities of stainless steel have made it the most widely utilised material for structural construction from the beginning of time. The most common reasons for using stainless steel are its durability, strength, and dependability. One fascinating feature about stainless steel is that it is a composite material made up of several different metals. As a result, different stainless steel grades are produced by varying the amounts used. Grade 301 is the most widely available on the market. This is due to the fact that it is simple to weld, and its ductile nature makes it perfect for usage in a variety of applications such as handrails, roofing, architectural cladding, and drainage sections, among others.

Carbon Steel is a type of steel that has a carbon content. Carbon steel is renowned for its tensile strength and wear resistance. It can be found on the structural framework beams that are used in highway construction, among other places. Bridges, welded frames, trailer beds, and some hollow structural components are also constructed with this material as standard.

Because it is flexible, it does not bend. Furthermore, it has the ability to tolerate extreme weather conditions without cracking. Low carbon steel contains 0.25 percent carbon and has two yield points, making it a low-carbon steel. The first yield point is slightly higher than the second, while the second is on the lower end of the spectrum. Carbon steel is widely used because of its adaptability and inexpensive price, which makes it popular. As a result, this steel is employed in a variety of various applications.

Carbide steel is created by combining iron with carbon in the proper proportions. The severity of the category goes from extremely high to modest. However, the amount of carbon present in the metal has an effect on this.

In the production of high-strength wire and tools, high-carbon steel is employed, whereas ultra-high-carbon steel is used in the production of cast iron.

Low carbon steel is utilised in ornamental ironwork such as gates and railings because of its low carbon content. Structural steelwork is often made of medium carbon steel (MCS).

Reinforced steel (also known as rebar steel)
In the construction industry, rebar steel is known as reinforcing steelworks, and it is used as a tension device in reinforced masonry or concrete structures. Carbon steel is used in the production of this type of steel, although ridges are added to aid in mechanical anchoring.

The steel forces the concrete into compression, which is why it is referred to as reinforcing. Rebar steel is available in a variety of grades with varying chemical composition standards, tensile strength, yield strength, and elongation percentages, among other characteristics.

The use of this form of steel ensures endurance and provides resistance over a large region that other steel types are unable to reach with their resistance. Rebar steel is available in a variety of diameters, and its expansion potential makes it particularly well suited for compressive concrete applications. This sort of steel is simple to instal prior to pouring the concrete, and it has the added benefit of reducing breaking and cracking, which is a regular occurrence in the construction industry.

Structural Steel is a type of steel that is used in construction.
The cross-section of structural steel is used to create the steel. It does, however, adhere to industry norms in terms of chemical composition and mechanical qualities. Structural steel is available in a variety of shapes, including HSS shape, K-shape, Z-shape, I-Beam, T-beam, Rail Profile, Structural channel C-beam, and others. HSS shape is the most common.

According to the construction, that form of steel is sturdy, ductile, and can be shaped into virtually any shape you can imagine. It is possible to use structural steel right away. Unlike other metals, it is transported to the construction site. Another advantage of structural steel is that it has a high level of fire resistance. Structural steel is more environmentally friendly than other metals due to the fact that it is recyclable and has a low carbon footprint.

Alloy Steel is a type of steel that is alloyed with other metals.

Carbon steel is combined with an alloying ingredient to create alloy steel. The inclusion of alloying elements improves the mechanical qualities of this steel by increasing its strength.

A good example is the combination of steel and aluminium. However, when you combine steel with manganese, you create a very strong substance that can withstand a lot of pressure.

Steel Building Construction Comes in a Variety of Styles
Steel construction, often known as steel fabrication, can be divided into several types, including:

Construction with bolted steel
The fabrication industry encounters this problem when they fabricate finished steel items and then ship the components to the construction site. The parts are then bolted together after they have arrived. Because the majority of the work is completed in the workshops, it is the most popular and favoured steel building approach. The procedure also helps to expedite the construction process because all that is required is to lift and bolt the steel components into place, which reduces the amount of time spent on site. Pre-engineered structures, for example, are an outstanding illustration of this style of construction.

What Role Does Steel Play in the Construction Industry?
According to the World Steel Association, more than 1600 million tonnes of steel were manufactured worldwide in 2016. According to these figures, steel is becoming increasingly popular for use in a wide range of construction projects.

Steel is used in a variety of applications, including:

When constructing a structure, steel is employed in the internal fittings and fixtures, such as stairways, railings, and shelves.

Steel structural sections - Steel structural sections are used by the majority of contractors to provide a strong framework for the building. In the construction industry, it accounts for up to 25 percent of the steel utilised.

Reinforcement bars - Steel reinforcement bars are used to increase the stiffness and tensile strength of concrete. Steel is sometimes used as reinforcement bars because it adheres well to concrete, according to some. Furthermore, it is a reasonably priced and durable solution. Steel can also be used to create deep basements and foundations for buildings.

Metal sheets are also utilised in a variety of building materials, including interior walls, roofing, cladding, purlins, and insulating panels, amongst others.

Non-structural applications — Non-structural steel can be found in a variety of other applications, such as interior ducting and heating and cooling equipment.

Utilities - Steel may be used to generate electricity, transport water, and burn fuel. For example, steel is used in the distribution of water and natural gas through subterranean pipes. Steel is also used in the construction of pumping stations and power plants.

Steel is frequently utilised in transportation networks, including rail tracks, bridges, tunnels, and structures such as ports, fuelling stations, and train stations. Steel is employed as rebar in these cases, accounting for around 60% of the total. The remainder is used on a variety of surfaces, including rail rails, plates, and individual portions.

What is the process of steel construction?

The process of erecting a steel building entails constructing the steel components into a frame on site and connecting them using processes that are both safe and dependable. All of the building components are made in a controlled environment prior to being delivered to the construction site, as part of the prefabrication process of the construction process.

It is important to note that steel erection includes various tasks such as:

Assuring that the foundations are sound and suitable for the installation of the structure
Lifting and securing steel components in their final positions. Cranes are occasionally employed, however jacking is frequently employed to attach items where bolted connections are necessary.
By keeping an eye on the column bases, you can ensure that the structure is in alignment. The bases must be lined, and the columns must be plumbed before they can be used.
Tightening the bolted connections in order to keep the frame in place

Conclusion

Steel is a construction material that is versatile, durable, inexpensive, and environmentally friendly. You will discover that steel was used in the construction of the majority of the buildings. It could be made of rebar, carbon steel, structural steel, or any other kind of steel. The type of steel to be used is determined by the physical and mechanical qualities of the steel.

Steel is classified based on its density, elasticity, strength, hardness, melting point, and thermal conductivity, among other characteristics. Consult with experts and conduct thorough research to decide the best steel to use in building.

CHAPTER 11:- BUILDING MATERIALS :- TMT BARS

For building projects, thermo-mechanically treated (TMT) steel bars are favoured due to their increased strength, high ductility, workability, and malleability, among other characteristics. TMT bars, on the other hand, were originally intended to provide structural support during earthquakes and to withstand high temperatures, particularly during fires. TMT bars, on the other hand, are available in a variety of grades, each of which represents a different level of stiffness and strength.

TMT bars are all suitable for building, however the different steel grades available are better suited to different types of constructions. Fe 415 TMT bars are suitable for residential construction, Fe 500 TMT bars are suitable for commercial construction, Fe 550 TMT bars are suitable for bridges and other slightly large-scale constructions, and Fe 600 TMT bars are suitable for large industrial projects.

It is the purpose of this chapter to discuss the various grades of TMT bars in greater depth, including their yield strength, tensile strength, elongation, and the types of building projects that are most suitable for each of them.

TMT Bars are available in a variety of grades.
TMT bars are steel bars that have been subjected to a quick cooling procedure while still in their molten condition, before being formed. This provides the steel bars with a malleable core but a hard and solid surface on the outside. A film coating on the outside of their outer shell provides corrosion protection, which makes them excellent materials for use in architectural constructions such as bridges and high-rise residential buildings in humid environments.

Water is sprayed on the TMT bars at regular intervals in order to cool them quickly. These intervals, on the other hand, differ, resulting in different steel grades in the steel bars. These classes are distinguished by a variety of features, including varying levels of stiffness and yield strength of the reinforcing material.

The following are the different grades of TMT bars:

Fe 600
Fe 550
Fe 500
Fe 415
Fe 415
Fe 415

The Fe 415 bar is the lowest grade, and the highest grade is the Fe 600 bar. Fe is an abbreviation for iron.

Depending on the amount of yield stress that they can bear safely and without experiencing irreversible deformation, these bars are classified into different grades. When stress is applied to the bar within its yield point, it will deform elastically, but when the load is removed, the bar will return to its original form or shape. The less bendable the TMT bar is, the higher the steel grade used to make it is.

It is expressed in Newtons per square millimetre squared (N/mm2), which is the yield strength. For example, a Fe 550-grade TMT bar has a yield strength of 550 N/mm2 and is used in the construction of bridges. The bar may be unable to keep its shape if the limit is exceeded beyond that.

Fe 415 is an alloy of iron and phosphorus.
The Fe 415 are TMT bars having a yield strength of 415 N/mm2 and an ultimate tensile strength of 485 N/mm2 that are used in the construction industry. Its elongation is 14.5 percent of its whole length.

As a result of their high uniform elongation, these steel bars are commonly employed in the construction of houses and other residential structures, as well as small-scale construction projects. Because of this property, they are also resistant to earthquakes. As a result, they are ideal for use in the construction of dwellings in seismically active areas.

These TMT bars also have a protective coating applied to them, which makes them more resistant to corrosion and rust. It can also be bent into the most intricate shapes with relative ease.

Fe 500 is a metal that is used to make a number of different things.
It has a minimum yield strength of 500 N/mm2 and an ultimate tensile strength of 545 N/mm2 and is made of titanium alloy. Its elongation is 12 percent of its original length.

The Fe 500 is referred to as a market-standard since it may be used in a wide range of construction projects and is widely available. In the construction of commercial constructions with numerous floors, smaller bridges, and underground structures, this bar has been used. It provides greater stability to high-rise construction projects while also allowing them to withstand greater loads. Aside from that, this grade of TMT bar is often utilised in coastal areas due to its anti-corrosion qualities, which makes it a popular choice.

Fe 550 is a steel alloy with a 550 tensile strength.

550 N/mm2 for the minimum yield strength and 585 N/mm2 for the ultimate tensile strength of Fe 550 grade TMT bars, respectively. Ten percent of the length is elongated in this product.

Given their higher tensile strength than Fe 500, these steel bars are suitable for a wide range of large-format, high-load capacity and industrial infrastructure development projects such as heavy bridges and underground constructions.

These TMT bars have a particularly negative influence on marine and coastal habitats as well.

Fe 600 Fe 600 TMT bars have a yield strength of 600 N/mm2 and an ultimate tensile strength of 660 N/mm2 and are available in a variety of sizes. Their elongation is approximately 10%. These steel bars are the strongest steel bars now available on the market, and they are utilised for heavy-duty industrial building projects involving enormous loads and long spans. Metro projects, expressways, buildings, plants, industrial zones, and massive commercial properties are all examples of such construction projects, as are others.

Why Should You Use TMT Bars in Your Construction Project?
A large amount of cold-twisted deformed bars were utilised for reinforcement in the building industry until TMT bars came into the picture and took over the market. As a result of a number of issues with CTD technology, including weaker ductility, higher carbon content, poor weldability, and increased corrosion rate, TMT technology has effectively taken over the low-cost reinforcement bar business.

Note, however, that not all TMT bars are made equal, and some are more expensive than others. Even yet, quality differs from one manufacturer to the next. When searching for the highest-quality TMT bars, it is important to consider the following characteristics:

They have the ideal blend of flexibility and strength, despite the fact that they have a low carbon content. When compared to CTD bars of the same grade, they have better yield strength, higher ultimate tensile strength, and a higher elongation %.
They are extremely weldable, allowing you to use them for a wide range of applications and applications.

They are extremely durable in the face of earthquakes and rust.

They include a significantly lower concentration of sulphur, which helps to make the structures more fire resistant.

How to Select the Proper TMT Bar

According to Build Supply, the availability of high-quality brands of TMT bars is the most important factor to consider when obtaining the appropriate TMT bars. Top-of-the-line TMT bars are available from reputable brands at competitive prices.

Furthermore, these businesses place a strong emphasis on methods for international-standard testing and monitoring in order to ensure that they supply consistent materials for your building projects. They are also completely clear in terms of cost, which allows you to make a decision in a more streamlined and expeditious manner.

In addition, according to Shyam Steel, the best TMT bars can defend a structure from fire, absorb additional energy in the event of an earthquake, and resist corrosion.

Furthermore, TMT bars have undergone a number of tests to ensure that they are of high quality in terms of their numerous qualities.

Flexibility and bendability: Because these are high-strength bars, they should have undergone a flex and re-bend test before being installed. TMT bars are bent to 135 degrees in this test, and they are then submerged in boiling water for approximately half an hour after being bent. After that, it is re-bent to approximately 157.5 degrees. In order for bars to bend readily, they must be produced with the appropriate combination of strength and flexibility. Cracks on the surface of the bar should not appear along the bends.

Resistance to compression: The resistance to compression of the bars is tested using a Universal Testing Machine. A TMT bar is placed in between the UTM's plates and pressure is applied to ensure that it does not break before reaching the yield point of the material.

Chemical composition: The chemical composition of the TMT bars must be in accordance with the IS Specification (International Standard). In order to produce steel bars of the appropriate grade, the chemical makeup of the steel must be precisely perfect. This chemical composition is measured with the help of a spectrometer.

CONCLUSION

The use of proper TMT bars is essential for any construction project, regardless of the needs and specifications of the project's core structure, or the quality standards that are being applied to it. To ensure exceptional quality, not only should you choose TMT bars from the top producers, but you should also ensure that you purchase the appropriate grade of TMT bars for your specific building project.

CHAPTER 12:- BUILDING MATERIAL :- CEMENT

The choice of the best cement for construction is critical to the structural integrity of the building. Varied varieties also have particular advantages and disadvantages that make them more or less suitable for different pursuits. However, the dispute remains as to which cement is the most effective for construction.

Ordinary Portland Cement (OPC) and Portland Pozzolana Cement (PPC) are two forms of cement that are suitable for use in building (PPC). OPC is available in three grades: 33 Grade for non-RCC projects, 43 Grade for plastering projects, and 53 Grade for projects that must be completed quickly. PPC makes the structure denser, making it ideal for large-scale concrete construction projects.

There are a variety of different types of cement that may be used for a variety of different projects, and we'll go through each one in great depth. By the end of this essay, you will have a better grasp of these types and will be able to select the most appropriate one for your project needs.

Ordinary Portland Cement and Portland Pozzolana Cement are two types of Portland Cement.

Almost every construction project necessitates the use of cement, but the type of cement that might be used will vary based on the project at hand. Most structures, on the other hand, must make a choice between two cement types: OPC and PPC. Both have advantages and disadvantages that make them appropriate for a variety of tasks. Understanding the differences between different types of cement is the first step in selecting the ideal one for your project.

A Quick Overview of Ordinary Portland Cement
Ordinary Portland Cement is a mixture of 95 percent cement clinkers and 5 percent gypsum, which is an addition that we use to speed up the setting time of the cement. It is more commonly used in building projects since it may be used for rapid tasks that do not necessitate the use of a range of concrete types.

However, because OPC is available in three different kinds, there are several aspects to take into consideration:

It is used for typical civil construction projects that do not require high-density concrete or RCC (Reinforced Cement Concrete) and do not require high-density concrete. After 28 days, it can acquire a minimum compressive strength of 33 megapascals (MPa). However, due to the needs of most buildings, as well as the availability of 43 and 53 Grade Cement, the use of 33 Grade Cement has become relatively outdated.

43 Grade Cement is the type of cement that we use for wall plastering, non-RCC constructions, paths, and other projects that do not require a high level of concrete density, such as foundations. During the first seven days of setting, it gains 23 MPa and reaches 43 MPa after 28 days of setting. It maintains its shape over time and is less capable of supporting as much weight, making it less suitable for RCC work.

Because of its increased density, 53 Grade Cement is suited for use in reinforced concrete construction, pre-stressed concrete, and mortars. 53 Grade Cement has a compressive strength of 27 MPa after seven days of setting, and it will eventually achieve a range of 53 to 70 MPa. 53 Grade Cement is typically used in any building project that necessitates stability and strength since it accumulates sufficient density over time.

The disadvantage of using OPC is that it does not provide the level of corrosion resistance and impermeability that some constructions might require. Despite the fact that both 43 Grade and 53 Grade are good for projects with a short turnaround time, they are not as long-lasting as what you would get from PPC.

A Quick Overview of Portland Pozzolana Cement
Portland Pozzolana Cement is a blend of 65 percent to 75 percent cement clinker, 15 percent to 20 percent fly ash, and 3 percent to 6 percent gypsum. Portland Pozzolana Cement is used in the production of Portland cement. Fly ash is a less expensive product that is similar to Portland cement in appearance. You will obtain a lower-cost structure when you utilise PPC for construction. This is perfect for buildings with a limited budget and that do not require as much strength as what you would receive from OPC.

PPC can be used for a variety of projects, including RCC and mass concrete projects, however residential constructions are the most typical application. Aside from the lower cost, PPC is a superior choice since it has a higher degree of fineness, impermeability, corrosion-resistance, and chemical-resistance—all of which are important attributes when building a house.

Despite the fact that PPC takes longer to set than OPC, it rises in strength over time. Because of the lower cost of small-scale construction, it is a more cost-effective choice for those with a restricted budget but plenty of time on their hands. PPC, on the other hand, may not provide you with the density you require if your building is made of high-strength materials.

Is it possible to use the same cement for all types of construction?

It is possible to utilise PPC and OPC interchangeably if you have greater quality control and adhere to best building procedures in your project. Although it is not always necessary, it is often preferable to use different types of cement in different parts of the building, particularly the foundation, plastering, and masonry work. It will allow you to be more cost-effective while still achieving the structural strength that you require for the project.

Aside from OPC and PPC, there are a variety of additional forms of cement that can be used in a variety of different structures. A appropriate cement should be used when working on a specialised building because of the different properties of each type of construction material. Consequently, your expertise with the numerous solutions accessible will be critical to the success of your project.

Other Types of Cement That Can Be Used in the Construction Industry
Cement is used in practically every type of building, and the two types we covered above are the most prevalent that you will come across. Because OPC and PPC cannot be used for all building projects, it is best to learn about the other forms of concrete that may be used for a variety of constructions, such as the ones listed below:

For projects in coastal areas, sewage treatment plants, and other water treatment facilities, Portland Slag Cement (PSC) is the cement of choice. Due to its increased resistance to corrosion compared to OPC, PSC is an excellent choice for structures that deal with sulphate and chloride, as well as places that receive a lot of seawater.

Resistant to Sulfate Structures that are subjected to sulphate assault require Portland Cement (SRC), which is the type of cement we utilise. Because it has better sulphate resistance than PCS, many sewers and water treatment facilities prefer to use it instead of the alternative. Using SRC in situations where chloride is present, on the other hand, could be hazardous.

Super Sulfate Cement is the type of cement that we use for mass concrete, particularly in constructions that are subjected to a great deal of chemical attack. It cannot be combined with OPC in order to increase strength because doing so will diminish the corrosion-resistant quality of the material. Aside from that, it's a very adaptable cement that expands when cured underwater and contracts when cured in air.

A quick-setting cement is the type of cement that is used for structures that are exposed to rushing or stagnant water. It has a composition that is comparable to that of OPC, but with a lower gypsum content and a smaller proportion of aluminium sulphate. With this combination, the cement hardens in 5 minutes and sets in another 30 minutes, depending on the temperature.

Rapid Hardening Cement is the type of cement that we use for construction projects that demand us to work at a rapid pace. It takes less time to develop strength, making it ideal for quick repairs and rehabilitation of structures in order to keep them in good condition.

Hydrophobic Portland Cement is the type of cement that is used in the construction of constructions in places that receive a lot of rainfall. An electrochemical coating operation is performed on it, which gives the building with water-repellent properties. Additionally, because portions of these structures are permanently buried underwater, it is the material of choice for the construction of tanks, reservoirs, dams, piers, bridges, and swimming pools, among other things.

Our most commonly used type of cement is white or coloured cement, which is used for decorative purposes. This type of concrete is unsuitable for most construction projects because it lacks the density necessary to support the weight of the structure. This product is typically used to fill in gaps between wall tiles, ceramic bathroom fittings, and wall décor. Although it has a more natural appearance, the manufacturing process for this form of cement is more sophisticated than that of OPC.

Conclusion
Modern construction cannot be completed without the use of cement. However, as we have a greater understanding of the structures that we might construct, we become more conscious of the requirement for various cement types that we should employ. In the past, we were limited to employing just OPC and PPC for practically every building project, but technological improvements have made better alternatives available.

CHAPTER 13:- BUILDING MATERIALS :- FERROCK

Ferrock is a carbon-negative, environmentally beneficial substance that can be used as an alternative to concrete in a variety of applications. It is made up of 95 percent recycled materials and is quite inexpensive to create, but how exactly is it made is a mystery.

In order to create ferrock, steel dust waste and silica must be combined. Iron carbonate is formed as a result of a chemical interaction between the iron in the dust and the carbon dioxide in the surrounding air. In the presence of carbon dioxide, it hardens into a solid form that is similar to concrete when mixed with water.

Ferrock is formed through a method that is described in this chapter. We will also cover the history of Ferrock and its prospective applications. We'll also discuss the advantages and disadvantages of utilising Ferrock.

What Exactly Is Ferrock?

It is possible to create Ferrock from mostly recycled materials, which is an iron-rich substance. When the materials are combined, they undergo a chemical reaction that results in a strong, hard solid form, such as concrete, as a result.

Ferrock is five times more durable than Portland concrete, according to the manufacturer. Furthermore, it is sufficiently flexible to avoid collapsing as a result of compression or seismic activity.

Concrete is a substance that we are all familiar with because it is utilised in so many different applications. This material is robust enough to be used in the construction of bridges and buildings hundreds of feet in height. Once cured, concrete can last for many years without deteriorating or becoming damaged in any way.

Concrete, on the other hand, has flaws that make it a poor choice for some projects, even when there is no other option available.

Saltwater is one example of a situation where concrete may not perform as well as it should, as the cement can corrode over time. Alternatively, sewage pipes, where repeated exposure to the chemicals used in water treatment would induce erosion, are another example.

The problem with making such a large amount of concrete is that the process releases carbon dioxide into the atmosphere. The manufacture of concrete and cement accounts for 5% of global CO_2 emissions, resulting in a significant amount of environmental pollution.

The manufacture of Ferrock releases carbon dioxide into the atmosphere, but the fascinating thing about this substance is that it absorbs CO_2 as it hardens. The greater the amount of CO_2 it absorbs, the stronger it becomes. As a result, Ferrock can aid in the reduction of greenhouse emissions, whereas concrete contributes to their production.

What is the process of making Ferrock?

Ferroconcrete is created by the use of steel dust waste products generated by a variety of different industrial operations. Typically, this dust is discarded, which is a waste of its potential. Heavy metals accumulate in landfills as a result of this practise, and toxins might seep into the earth as a result.

To make silica, you'll need to use glass that has been ground up in addition to the steel dust waste. When you acquire large products such as bags, jackets, and other large items, you may find this product in the shape of a small paper packet in the packaging.

Ferrock is made by combining steel dust and silica with ferrous rock, which is an iron-rich mineral, and other components that produce corrosion or rusting to form a cohesive mixture. After that, water is added to the concoction, which transforms it into a paste that resembles the chalky mess of cement.

Once the Ferrock has been mixed, it can be used in the same way as cement is. You can use it to pour into any mould or to trowel it with. At the end of the process, the mixture is exposed to carbon dioxide gas.

As a result of the CO_2 absorption by the iron in the dust, the iron in the dust will begin to rust. The carbon dioxide is absorbed into the mixture and fuses with it to generate iron carbonate. Ferrock is the solid version of the substance that hardens after a week of exposure to air.

Because of the way the Ferrock hardens, it is able to trap carbon dioxide within the material. Surprisingly, the addition of CO_2 actually helps to strengthen the Ferrock's structural integrity. The consistency of Ferrock is similar to that of concrete, in that it cannot be reverted to a liquid form once it has set.

Ferrock has a number of advantages.
In comparison to concrete, there are numerous advantages of employing Ferrock as a building material in construction. We'll take a deeper look at these advantages, which include strength, flexibility, carbon neutrality, and the fact that it's non-reactive to chemicals.

Five times more difficult to break than concrete

Ferrock hardens into a solid mass that is comparable to concrete in appearance. Ferrock, on the other hand, has five times the strength of concrete once it has solidified. This means that it can withstand a greater amount of weight, compression, and damage without becoming damaged.

The compressive strength of this material ranges from 5,000 to 7,000 psi. Some experiments reached pressures of up to 10,000 psi. These pressures are higher than the approved OPC standard values for concrete, which are 4786 psi (OPC-33 MPa), 6236 psi (OPC-43), and 7687 psi (OPC-53) respectively.

Ferrock's durability and stability make it suitable for use in the same applications as cement, such as the production of construction materials. Moreover, when made into foam, this material can endure temperatures of over 1000°F (600°C), making it a great contender for fireproofing or insulating applications.

Flexible
Ferrock is also somewhat flexible, which means it can withstand greater pressure and movement than other types of rock. Because concrete is fully solid, even the smallest movement can develop fissures in the structure, which can cause the entire structure to collapse.

However, Ferrock is capable of withstanding some movement without being compromised. When used in places with active seismic activity, such as earthquake-prone areas, this feature is very advantageous. Ferrock was shown to be four times stronger than Portland cement in flexural tests, according to the researchers.

Inactive in terms of CHEMICALS

Ferrock is considered chemically inactive, which implies that it does not decompose when exposed to gases or chemicals, as opposed to other materials. Concrete can deteriorate as a result of exposure to chemicals and the passage of time.

It is for this reason that Ferrock is frequently used in marine building, as it is impervious to the effects of saltwater. The truth is that when exposed to seawater, Ferrock actually becomes more resilient, making it a perfect choice for underground conditions.

It is also resistant to a variety of circumstances such as ultraviolet radiation, corrosion, rotting, rust, and oxidation, among others. It is also resistant to chemical damage, making it a desirable solution for pipes and tubes of many kinds.

Neutral in terms of carbon dioxide emissions

Ferrock is also considered carbon-neutral, which indicates that throughout the production process, it does not release a significant amount of carbon dioxide. During the manufacturing process, this material emits a little amount of CO_2.

Ferrock, on the other hand, while in the liquid state, makes use of carbon dioxide to aid in the hardening process. CO_2 fuses with the other elements in the combination, trapping the gas inside the rock as it solidifies.

As a result, Ferrock acts as a carbon dioxide filter, eliminating a portion of the CO_2 that is present in the environment. It uses the CO_2 it has absorbed to build its final shape, which is a solid sheet of hard Ferrock.

Ferrock has a number of disadvantages.
By this point, you've most likely come around to the notion of using Ferrock in place of concrete in your project. We're right there with you in every step. However, there are significant disadvantages to using this material over an extended period of time that make it inappropriate for this purpose.

Materials

Ferrock is a concrete substitute that is more environmentally friendly. Ferrock is composed primarily of recycled materials, which means that you will not be depleting natural resources in the same way that you would with concrete construction.

However, this is where the difficulty lies. Both the steel dust waste and silica required for ferrock production are byproducts or leftover scraps from a different manufacturing process. As a result, there is a limited availability of both of these products.

Ferrock's production is heavily reliant on the production of other items. What this means for Ferrock makers is that the amount of rock they can produce is limited by the availability of the raw materials.

Ferrock is made up of a lot of silica and metal shavings, which makes it difficult to use it for large-scale construction projects. Current application possibilities for what you can build using Ferrock are somewhat limited.

Costs

Because Ferrock hasn't gained widespread popularity yet, it's still rather easy to come by the stuff you'll need. As a result, you can make Ferrock for a relatively cheap over-cost. However, because of the production process, concrete is usually less expensive to produce or acquire than Ferrock.

However, if businesses begin to discover that there is a profit to be produced from their garbage, expenses may rise, and it may become more difficult to locate the resources required.

Over time, it is possible that using Ferrock instead of concrete will become prohibitively expensive, notwithstanding the growing environmental concerns associated with the release of greenhouse gases during the concrete manufacturing process.

Untested
One of the most major obstacles preventing Ferrock from completely replacing concrete is the fact that it is still in its early stages. The use of concrete has been around for more than 200 years, and there is still a lot that we don't know about it.

Ferrock, on the other hand, has proven to be strong enough to bear greater compression and weight than concrete. However, because it has only been in use since the early 2000s, we do not know how long the material's life cycle is expected to last.

Ferrock is also unknown in terms of how well it will perform in relation to building conditions, as it requires specific hardening processes to be effective. On top of that, it's unknown whether the same types of concrete processes can be used on Ferrock.

Ferrock's illustrious past

Doctor David Stone, creator and owner of Iron Shell Media Technologies, as well as a former Ph.D. student in the Department of Soil, Water, and Environmental Science at the University of Arizona, is credited with inventing Ferrock.

Ferrock was created by accident by Stone while he was working on a project back in the year 2002. At the time, he was looking at techniques to keep iron from rusting and hardening as much as possible.

At first, he wasn't impressed with the material he had made and decided to abandon the experiment. However, he soon changed his mind and chose to concentrate his efforts on developing a substance that has the same physical properties as concrete, but in an environmentally benign form.

In order to put his novel notion to the test, he collaborated with the Tohono O'odham Nation Reservation in Southern Arizona to obtain the silica that he needed for his experiments.

In addition, he got subsidies from the Environmental Protection Agency (EPA) totaling $200,000, which enabled him to develop demonstration projects with the help of the Native American community.

The Ferrock production process was figured out, and Stone entered his environmentally friendly alternative to concrete in a competition, which he was successful in winning.

The innovation was granted a patent by the United States Patent and Trademark Office in 2013. Despite the fact that Stone is the one who invented Ferrock, the University of Arizona owns the copyright on it because he was working for them at the time of his discovery.

Stone worked out a contract to hold a licence for his innovation a year later, in 2014, which allowed him to market his technology. This licence was obtained in partnership with the Arizona Tech Launch Center (TLA)

Iron Shell Material Technologies are a type of material that is made of iron shells.
Ferrock is currently protected by a patent held by the University of Arizona. Iron Shell Material Technologies has been granted an exclusive licence by the company. Ferrock can now be manufactured and sold as a result of this licence, which allows it to be used for commercial purposes. Ferrock has now been registered as a trademark.

In order for Dr. Stone to be able to sell his invention, he established Iron Shell Corporation. His company is dedicated to the discovery of novel iron-based materials that are environmentally beneficial, carbon negative, and may be used in building.

Given the company's mission, it should come as no surprise that Stone employs steel dust that he obtains for free from steel companies that do not participate in the recycling of leftover steel pchapters. Thus, heavy metal pollutants are not dumped into landfills, and the environment is protected.

He also spends his spare time collecting empty bottles from the Tohono O'odham Nation reservation, which he sells. Stone doesn't just collect the leftover bottles from the drink dispensaries with the help of tribal member Richard Pablo and others; he also gets them from the liquor stores.

Dr. Stone also contributes to the clean-up effort by collecting up bottles that have been left on the side of the road. Afterwards, the bottles are fed through a glass crusher, where they are crushed and transformed into silica.

Polymiron
Iron Shell has also developed polymiron, a carbon-negative ionomer that contains a high concentration of iron, in addition to the invention and patenting of Ferrock. This biopolymer, Dr. Stone and his colleagues think, will be able to be used as a sealant, tar, glue, and other applications in the future.

Because this product is electrically conductive, there is a possibility that it will be used in energy conversion systems. This product, on the other hand, is still in the development stage.

Ferrock has a variety of applications.
Ferrock is currently being utilised as a substitute for a variety of modest projects, such as pavers, tile, and bricks, among others. It has also been tested and certified for use on larger items such as slabs, sidewalks, walls, and benches, among other things.

On the other hand, there haven't been any substantial developments in terms of large structures, roads, or any potential projects that might arise in the future.

There have been some encouraging results, which open the door for Ferrock to enter the marine-based construction industry. Due to its remarkable capacity to become more resistant when exposed to seawater, rather than decaying like concrete, which is simply clay and chalk combined together, it is becoming increasingly popular.

Additionally, ferrock can be utilised for marine structures such as structural pilings, seawalls, piers, breakwaters, and foundations, among other things. It is also suitable for use in the construction of piping since it is resistant to pollutants found in sewage, such as sulfuric acid, and is therefore cost-effective.

What are the advantages of Ferrock over concrete?
A proven and true commodity with a long history of use, concrete has been around for hundreds of years. We've figured out what we're going to do with it. And it can be found everywhere you look.

Nonetheless, despite the widespread usage of concrete, there is a compelling reason for us to explore alternative building materials in the future.

A tonne of carbon dioxide is released into the atmosphere during the manufacturing of concrete and cement – which is created separately and then added to the concrete to firm it up and bind together to form a hard shape – each year.

The reason for the high levels of CO2 is that cement must be heated to extremely high temperatures in order to break down the limestone in order for it to be effective. High, as in 2,800 degrees Fahrenheit (1537.778 degrees Celsius), is what we mean by scorching.

To give you an idea of the amount of carbon dioxide we're talking about, for every 1000 kilogrammes of cement produced, 900 kilos of carbon dioxide is emitted into the atmosphere by the cement plants themselves.

Allow me to introduce you to Ferrock, which has been discovered to absorb carbon dioxide and fuse it into the matrix. It appears to be a no-brainer that we would want to utilise more of a product that can perform the same functions as concrete while also purifying the air in our surroundings.

Decide on a time frame
Another advantage of Ferrock over concrete is that it does not require as much time to set and harden as concrete does. It takes 24 to 48 hours for concrete to harden. However, it can take up to 28 days before it achieves its maximum potency. Hard impacts to the concrete might still cause damage to it prior to this time frame.

Ferrock will require at least a week to complete its expansion. Over the course of this period, all of the small glass shards fuse together, and the material becomes harder. It also continues to draw in carbon dioxide, which contributes to the durability of the device by increasing its capacity.

Conclusion

Ferrock appears to be a realistic answer for a carbon-negative, recyclable, and long-lasting material that could be used to replace concrete in some applications. The fact that there are people out there attempting to figure out ways to lessen our carbon footprint is encouraging, especially in light of the fluctuations in global warming over the last few hundred years.

And what better way to accomplish this than by utilising an environmentally friendly material created from recycled materials? In addition, the fact that Ferrock has the ability to absorb carbon dioxide from the environment, so lowering greenhouse gas emissions, makes it a superior product.

CHAPTER 14:- BUILDING MATERIALS :- AIRCRETE

Many builders in the building business today believe that aircrete (autoclaved aerated concrete) is the way of the future; however, there are concerns about the quality and longevity of this material in some cases. Many homeowners have inquired as to whether aircrete is fireproof in order to reduce any potential future risks.

Aircrete is fireproof because it is constructed of fireproof components, such as a cement mixture, foam, and water, which are combined to form a solid structure. No matter how much heat is applied to aircrete, it will not catch fire. If you're looking for a high-quality, fire-resistant building material, aircrete is a great option to consider.

For a variety of reasons, the construction and design of this product make it the most suitable option. Aircrete can be used in a number of construction projects, ranging from minor house renovations to large-scale high-rise construction. Continue reading to learn more about how and why aircrete is beneficial.

So, what exactly is Aircrete, and how does it come to be used?
Aircrete is the result of the combination of numerous ingredients, including calcium gypsum, quartz sand, aluminium powder, lime, cement, and water. Calcium gypsum is the primary component. The mixture is then cured in an autoclave to form blocks once it has been blended with the other ingredients. In addition to offering excellent insulation, these blocks are resistant to fire and mould as well.

When the aircrete mixture is placed in an autoclave, a chemical reaction takes place that confers key properties on it, such as fire resistance and waterproofing. When the aluminium powder combines with the calcium hydroxide, hydrogen fumes are produced as a byproduct. Once the gas bubbles have been expelled, the air is restored, which helps to preserve the little air bubbles that are characteristic of aircrete.

Building with aircrete is not a new sort of construction material. In reality, it has been in use in the building industry for more than seven decades. One of the most appealing aspects of this product is that it has a minimal influence on the environment. Because of its adaptability, it is a popular choice in a wide range of diverse construction settings, including the following:

Placing items both inside and outside
Buildings that are subjected to extreme temperature variations
Buildings with a lot of floors
Assembly of the interior of the cavity wall
Construction of a solid wall
Malls are places where people go to shop.
Airports \sHomes
A commercial building is a structure that is used for business purposes.
Insulation
A steel frame is shielded by a protective barrier.
This highly adaptable material is well-suited for any work, regardless of its size. In applications ranging from large-scale high-rises to small-scale household projects such as garden boxes, aircrete can deliver a robust product with characteristics similar to those of ordinary concrete.

Aircrete has a number of advantages.

When it comes to your next building project, there are various advantages to using aircrete. For numerous reasons, including its ease of use and fireproof properties, aircrete is a popular choice among contractors in a variety of circumstances. Not only is it equivalent to normal concrete material, but it also provides far greater flexibility, making it a superior alternative for a wide range of construction applications.

Material that is lightweight
Because of the manufacturing technique used to create aircrete products, their lightweight construction makes them excellent for a wide range of applications. The use of adaptable blocks that are simple to handle, shape, and drill can greatly reduce the amount of time it takes to finish a building project overall. Saving time on the job site will not only save money by reducing labour costs, but it will also lower the prices of supplies delivery, which will lower overall costs.

Using enormous blocks of prefabricated aircrete, labourers may execute big-scale construction projects more quickly and efficiently. This is evident in the reduced amount of masonry work required, as well as the fact that these lightweight aircrete blocks are easier to cut or mould as needed, as opposed to typical concrete materials.

Fireproof Aircrete is distinguished by its remarkable ability to not only resist fire but also to shield and prevent other areas from being damaged by fire. This fire-retardant property is due to the porous nature of the material, which makes it a desirable choice for a wide range of applications and environments. They are designated as a non-combustible material and adhere to all applicable construction norms and regulations.

Aircrete has numerous advantages regardless of how it is used, whether as an insulation material or as a firewall. Electrical fires that have started in homes that have used conventional insulating measures have resulted in widespread devastation and death. Aircrete has the ability to alter the result of a wide range of extreme conditions, like restricting the spread of fire from within walls or even from a nearby building.

Waterproof and pest proof are two characteristics that distinguish this product.
A major source of concern for building owners is the presence of water and water damage. Because of the way it is constructed, aircrete provides excellent ventilation.

By allowing for the dispersion of water, aircrete contributes to the reduction of humidity in a structure by absorbing and distributing the moisture that accumulates there. This procedure can dramatically reduce the likelihood of mould or mildew forming in the home or business. The chemical composition of aircrete is such that it will not breakdown, decompose, or rot when exposed to water.

When attempting to keep pests at bay, aircrete can be a feasible option to consider. These lightweight blocks close up any openings that pests might use to gain access to a building. Aircrete seals effortlessly along the perimeter of the floor, as well as around doors and window frames, preventing any possible problems caused by little intruders. In addition, they do not attract pests and are therefore invulnerable to rodents and other external pests.

It is long-lasting and durable.
Its long-term endurance makes aircrete a superior building material in comparison to traditional uses such as regular concrete. A significant benefit for climates around the world is the capacity of buildings to resist extreme temperature variations without sustaining damage or altering their original design and construction. Its primary design goal is to withstand the degradation that occurs over time as a result of ordinary environmental conditions.

Even in the most extreme conditions, Aircrete will outlive the environment. Years of use in a variety of applications have demonstrated that this autoclaved aerated concrete material outperforms numerous traditional processes in almost every manner. Aerial concrete materials might be beneficial for buildings that are subjected to natural disasters such as earthquakes, tornadoes, and hurricanes.

Economical
When properly placed, aircrete has the capacity to soundproof a building, which is a feature that is much appreciated. This autoclaved aerated concrete material, when properly installed between floors and walls, can aid to reduce noise in the construction environment.

If a building is properly insulated, it will retain heat and maintain a comfortable temperature when necessary. Aircrete can assist in lowering heating and cooling expenses for buildings by assisting in the regulation of the temperature inside the building. It is non-toxic, and there are no toxins used in the autoclave process, which means that it will not pollute the finished building's environment.

Aircrete is also noted for being environmentally friendly in a number of different ways. It contributes to having a lower overall environmental effect during the manufacturing process and the transportation of materials on a job site. Alternatively, the use of seamless integration during the installation of aircrete helps to reduce the amount of trash produced.

Aircrete has a number of disadvantages.
Despite the fact that there aren't many drawbacks to using this building material, you should be aware of a few potential concerns before proceeding.

Aircrete can be brittle when handled before to installation, and therefore must be handled with greater care than ordinary bricks in order to avoid any damage to the blocks during the process. Given the fragile nature of these autoclaved aerated concrete blocks, it is advised that thinner and longer screws be used for any attachments to prevent breakage.

In addition, aircrete should not be erected in wet conditions because it has been seen to crack and sustain damage when wet. All of the blocks should be completely dry before being installed and subsequently. Aircrete should be handled and installed in accordance with the manufacturer's recommendations in order to minimise any potentially serious side effects.

Conclusion
Despite the fact that aircrete has been around for many years, the number of applications where it is preferred is rapidly rising. As more contractors become aware of the numerous advantages of this autoclaved aerated concrete product, its popularity continues to grow. The adaptability of aircrete, which can be used for everything from insulation to walls, floors, and ceilings, provides it an advantage over traditional approaches.

The ability of Aircrete to withstand fire, in addition to its many other characteristics, ensures that structures meet the specified requirements required for proper construction. If you're looking for a cost-effective solution that will help you save money on your heating and cooling bills while also repelling pests and keeping your family safe, aircrete is the product for you.

CHAPTER 15:- BUILDING MATERIALS FOR HUMID CLIMATE

If you live in a humid climate, you are well aware of the difficulties that can arise. With all of the moisture in the air, it's possible that some of your efforts will come to a grinding halt. Have you been on the lookout for the greatest building materials to use in your humid environment? Engineering hardwood, spray foam insulation, clay or lime plasters, adequate windows and ventilation, waterproof paint and sealants are the ideal materials for humid regions. Concrete can be used in humid locations, but it must be combined with efficient humidity management to be effective. Take a deeper look at what causes a humid climate before we get into some of the greatest building material options now available on the market.

What Contributes to a Humid Climate?

Is it possible that you've been out with your friends or family and one or more of them has commented on how humid it is? Perhaps you have made similar remarks in the past yourself. Has it occurred to you to question what it is that makes the surroundings of humid climates so unbearably unpleasant? It's a straightforward response, but there's a lot of science behind it as well. The simplest explanation is that a humid climate has a high concentration of water vapour in the atmosphere. The greater the amount of water vapour in the air, the more humid the air will be. As water evaporates from the earth's surface, this water vapour is released into the atmosphere. Because heat causes water to evaporate more quickly, humid areas tend to have higher levels of humidity. Humidity will be particularly high in areas with a significant concentration of trees. This is caused by the amount of moisture released back into the atmosphere by trees through a process known as transpiration. Essentially, this is water that has been absorbed by the soil and is being released by the leaves.

How to Select the Most Appropriate Materials

Choosing materials that are durable in all climates will be essential if you're constructing any kind of structural project. These buildings will be subjected to a great deal of abuse from the elements, including wind, rain, animals, cold weather, and heat. The materials you use must be able to resist the extreme conditions. Choosing outside siding is extremely important when it comes to a home's appearance. Your siding will be exposed to moisture, which is the most harmful element that it will encounter. Wood and clay are both excellent building materials for warmer regions, where humidity is more likely to be a problem. It is critical to use materials that are lightweight and have a low thermal mass in regions where the days are hot and humid and the nights are cool. The ability of a material to absorb and store heat is referred to as thermal mass. Because of the significant amount of heat energy required to change the temperature of concrete, bricks, and tile, these materials have a high thermal mass. In hot climates where the daytime temperature is high but the nighttime temperature is low, concrete is an excellent choice for construction. If you live in a humid region, using high thermal mass materials may help you avoid some moisture problems in your structure, but it will require a lot more energy to cool those materials down. When building your structure, walls, and ceilings, it is ideal to choose materials with low thermal mass, such as wood.

Humid climates need the use of the best building materials.
Because humidity is such a widespread problem that many various areas and climates must contend with, there are many different materials that may be used to combat it. The following is a list of the eight greatest building materials for areas with high humidity.

Hardwood that has been engineered

Wood is a typical building material, but due of its potential to absorb moisture, it may not be the ideal choice in many situations. If you live in a particularly humid region, engineered hardwood, on the other hand, may be a much better choice. Engineered hardwood is produced in high-temperature conditions, and it is formed by bonding a thin layer of wood to a pressure-treated plywood or fiberboard. This increases the durability and watertightness of the material, which helps to prevent warping and mould growth. As a result, it is an excellent choice for flooring in humid conditions. It is possible to purchase engineered hardwood flooring planks from sites like Home Depot, which offer a selection of species and stains to choose from.

USG Structo-Lite Basecoat Plaster is a high-performance basecoat plaster.

Natural plaster has the ability to absorb interior humidity and slowly release it over time, which makes it an excellent choice for bathrooms. Clay and lime plasters are naturally mould resistant, making them appropriate for use in humid locations where mould growth is a concern. The fact that clay is capable of enabling moisture to travel easily through the construction is what makes it such a desirable material for humid areas. Clay plaster is also capable of maintaining interior humidity levels by retaining moisture when levels reach more than 50% of its maximum capacity. Once the humidity level falls below 50%, the plaster will begin to release the moisture back into the atmosphere. USG Structo-Lite Basecoat Plaster is an example of a clay plaster that is suitable for use in humid areas, such as tropical regions. The only ingredient required for this plaster is water, and it weights significantly less than some of the competing products on the market.

Eagle Concrete is a company that produces concrete products.

Concrete may be an useful building material in a humid area if the structure is well ventilated and the humidity levels are maintained within the structure. A lack of air circulation and climate management will cause excessive moisture to build up in concrete, causing it to deteriorate. In a building where heat is allowed to enter through vents, windows, and other openings, condensation will form if there is insufficient circulation of the air. Once this condensation happens, it will condense on the concrete and have the ability to cause harm to the surface. The need of proper ventilation and airflow throughout the building, as well as sealing it to ensure it is completely airtight, can not be overstated. Eagle Concrete, which is available at Home Depot and is a water sealant concrete that repels water. Because the water condenses on the concrete, it is less likely to cause structural damage.

Insulation Made with Spray Foam

Open-cell and closed-cell spray foam insulation are the two types of spray foam insulation from which to pick. This will depend on where in the building you intend to use the spray foam and how much you want to spend on it. In humid climates, open-cell spray foam should be sprayed on attics, walls, and roofs to keep moisture out. As opposed to closed-cell foam, open-cell foam allows moisture to move through the building without inflicting damage to other parts of the structure. The use of closed-cell spray foam — or open-cell foam painted with a vapor-resistant paint — in basements and under floorboards is recommended because damage from external moisture is more likely to occur. Closed-cell spray foam is also an excellent choice for protecting portions of your structure that may be vulnerable to floods. Waterproofing spray foam from Home Depot, Touch-N-Foam spray foam, is a closed-cell spray foam that is moisture resistant. It also does not expand or contract, making it a long-lasting material to use for construction.

ThermaStar by Pella Windows is a high-efficiency window system.

Ventilation is essential for maintaining an acceptable amount of interior moisture. Because of the high moisture content, it is important to guarantee that the materials used inside, such as the joints and wall studs, do not mould or rot. The most effective method of obtaining natural ventilation is through large windows that allow a cross-breeze to circulate throughout the structure. The most appropriate type of window to use in humid conditions is an energy-efficient one that is well-insulated and sealed. The presence of moisture can enable hazardous mould and other microorganisms to grow if the area is not properly sealed off. The ThermaStar by Pella, which is available at Lowe's, is an excellent example of an energy-efficient window. This window is constructed of vinyl framing that has been designed to endure the elements.

Hugger 52-inch Indoor Ceiling Fan

In the event that opening windows is ineffective in achieving the desired airflow through the building, installing vents is another option. Bathroom vents and vents above the stove can both assist in reducing the amount of moisture in the air left over from showers or cooking activities. The use of ceiling fans is another excellent method of ensuring that ventilation is maintained throughout your facility. These fans not only help to keep the temperature down, but they also help to ensure that moisture in your structure evaporates more quickly. Ceiling fans, such as the Hugger 52in Indoor fan from Home Depot, are excellent choices that are not prohibitively expensive. The ability to maintain cold and dry air within the structure will help to keep the humidity at bay.

Rust-Oleum Zinsser PermaWhite Exterior Paint is a white exterior paint produced by Rust-Oleum Zinsser.

Because of its capacity to inhibit mould and mildew growth, this paint makes an ideal building material. Almost all paints have the ability to act as a moisture barrier, protecting your building from the elements. Rust-Oleum Zinsser exterior paint goes above and beyond the capabilities of conventional paint. It contains a powerful mildewcide that prevents mould from growing on the paint itself as a result of dampness.

Thompson's WaterSeal is a type of sealant.

Building moisture sealants are essential for ensuring that a structure is protected against the moisture that is common in humid conditions. Sealing wood is especially important due to its organic makeup. Wood rots as a result of moisture in the air, and the greater the amount of moisture in the air, the more quickly this could potentially occur. Making sure that wood is sealed will help prolong its life. WaterSeal, a product by Thompson's WaterSeal, is an example of this. It can be purchased on Amazon. It simply requires a single application and provides excellent protection against water and moisture. It can be used on decks, fences, and other outside wood.

CHAPTER 17:- BUILDING MATERIALS :- DRY UNDER WATER EVEN

Did you know that in order to construct bridges, construction crews must first pour concrete foundations into a river or the ocean's bottom? Concrete is made out of a mixture of sand or gravel, water, and cement, with cement serving as the binding agent that holds the material together. However, have you ever pondered why concrete does not become diluted or washed away when submerged in water?

When exposed to water, concrete dries faster and more thoroughly than when exposed to air. This occurs as a result of the hydration of cement pchapters. The cement reacts chemically with the water, forming a chemical bond that holds the sand and gravel together. This curing (hardening) process, which takes nearly a month, causes the concrete to harden and become set.

It is therefore necessary to keep your concrete moist during the curing process in order to gain the most strength and durability from it. Continue reading to find out more about how concrete sets underwater.

What is the process through which concrete sets under water?

When concrete is poured underwater, one of its constituents reacts with the water to generate an outer coating on the surface of the concrete. This covering prevents a large amount of water from either seeping in or, worse, from diluting the cement, which would otherwise occur. Another chemical in the concrete then reacts at a slower rate and sets to its final hardness 28 days after the first reaction occurs.

While the majority of people assume that the reason cement sets is because the water that has been mixed in evaporates, this is not entirely right. As previously stated, cement sets as a result of the chemical reaction that takes place. After that, the majority of the water in the mixture is consumed by reaction with compounds inside the cement, resulting in the formation of new compounds that are extremely difficult to work with.

When concrete hardens, it does not "dry" in the traditional sense since doing so would impair the chemical process. As a result, concrete that cures underwater tends to be stronger than concrete that cures in the open air.

However, this is not true for non-hydraulic forms of cement, which are those that use lime and gypsum plaster as a binder and whose curing process is dependent on water evaporation. Decorative cements such as these are beneficial for a variety of purposes such as historical restoration.

What is the best way to get concrete to set underwater?

Hydraulic cement is used in the construction of the majority of concrete structures, including buildings and bridges. This means that the hydraulic cement used in the majority of conventional concrete aggregate combinations serves as the binder. This is followed by the use of water, which allows for the chemical reaction that is the foundation of the setting process to take place.

As long as the necessary amount of water is maintained in the concrete mixture, the concrete will set over time.

So, what can you do to make sure that the drying process runs as easily as possible? When placing concrete, how can you ensure that it does not fall into the water?

One method of accomplishing this is through tremie or the use of a concrete pump. Additionally, you can prevent the unset mixture from spreading by employing certain concrete specifications or constructing a customised framework on top of a water bed.

The Best Way to Prevent Cement Washout Underwater is to...
One method of preventing cement washout while curing concrete underwater is to increase the setting time of the concrete mixture. This can be accomplished by including admixtures that will aid in the process. Alternatives to aggregate, portland cement, and water are used in the production of concrete, and they are added either before or during the mixing process.

They aid in the assurance of quality concrete during the mixing, transporting, placing, and curing operations, as well as the modification of the characteristics of hardened concrete.

How to Ensure the Longevity of Your Concrete Construction Projects

Concrete is one of the most durable and visually appealing building materials available, and its use is far higher than that of steel or wood. However, the strength of the mixture is determined by what you do after you pour it, just as much as it is by the mixing procedure you utilise.

It is critical to keep the concrete moist during the curing process. If the water evaporates from the surface of the product at an excessive rate, the final product will become weak as a result of tension and breaking. Specifically, this occurs when the mixing takes place outside, in the direct sunlight.

What happens after that? The first several days are really important. This is due to the fact that you must maintain control over the moisture content and temperature of the fresh concrete. Providing particular attention to the concrete mix throughout the curing process helps to increase the structural integrity of the finished concrete. This increases the material's resistance to cracking in the future.

Additional recommendations to help ensure that your concrete structures are sturdy and long-lasting are provided below:

Water should be sprayed on new concrete.

The process of hosing down concrete with water is a typical method of curing concrete. During the first week, you can do this as much as five to 10 times per day.

We call this process "wet curing," because it permits moisture contained inside the concrete to gently escape. Furthermore, concrete that has been cured using this procedure is nearly 50% stronger than concrete that has not been cured with moisture. This procedure, however, is not suggested for concrete that is being poured in cold weather conditions.

Make certain that the new concrete is protected.
If, for whatever reason, you are unable to spray your concrete with water, you can use a cover to retain and slow the rate of moisture evaporation in the mix by placing it over the concrete. Polyethylene sheeting or a curing blanket should be used. The thickness of the sheets should be at least 4mm (0.157 inches).

After fully saturating the concrete, cover it with the sheeting and fix it with something heavy, such as bricks or rocks, to prevent it from shifting. Take down and replace the sheeting every day for the next seven days, after which you will moisten the concrete and replace the sheeting. In addition to concrete columns and walls, you can utilise this technique on wood.

Pond Cure Concrete Slabs should be used.

This is yet another way for curing your concrete that you might employ if necessary. In order to accomplish this, you must first construct temporary berms around the new concrete slab before flooding the interior space with water up to a foot deep (30.48cm).

Pond curing is a three-day process that requires a lot of patience. Nothing needs to be done on a daily basis—just make sure that the water level stays above the concrete slab at all times.

It is important to note that forming berms around a large concrete slab, such as a foundation slab, requires a significant amount of dirt. As a result, large-scale construction companies frequently employ this strategy since it allows them to complete their projects more quickly. It enables them to pour the foundation slabs of the construction rapidly and move on to the framing phase.

Applying a Curing Compound Makes the Curing Process Easier

Finally, curative ingredients are a much more straightforward option. This is due to the fact that they include soluble emulsions, which form a protective covering on freshly poured concrete slabs or walls when applied. When sprayed directly onto the wall surface, this coating creates a protective film that stops water from evaporating from the surface. This permits the wall to cure at a consistent rate throughout the process.

While some curing ingredients totally disintegrate after a few weeks, others are washed away after the curing process is complete. Others, on the other hand, penetrate the concrete to form a permanent sealant that waterproofs the concrete while giving it the appearance of having been recently poured.

Conclusion

Consider drying your concrete underwater in order to produce a high-quality, crack-resistant final product. It should be noted that some concrete slabs may still crack as a result of concrete shrinkage that occurs during hydration — water is used up and the temperature fluctuates — and should be avoided.

To prolong the life of your concrete slab and keep it looking beautiful, take these steps:

Control joints should be placed at precise locations within one day after the pour. Use a metallic jointing tool to cut neatly into the concrete surface; the joints will act as a guide for the inevitable cracks that will appear.
It is important not to allow the new concrete to become too cold because this will interfere with the chemical hardening process. If the weather becomes chilly, use a concrete insulating blanket to keep the concrete warm.
Excessive weight should not be applied to new concrete. Concrete reaches its maximum strength in around 28 days, so avoid allowing foot movement on a freshly poured slab for the first 24 hours after it has been laid.

CHAPTER 16:- BUILDING MATERIALS :- BLACKTOP AND ASPHALT

The terms "blacktop" and "asphalt" are frequently used in the same sentence. It should be noted that there are some substantial distinctions between the two. This chapter will discuss the differences between the two types of pavement surfaces in order to dispel some of the misconceptions around these paving surfaces.

The method by which blacktop and asphalt are combined is the most fundamental distinction between the two. Despite the fact that they contain the same ingredients, they are prepared in a different way. Another significant distinction is the manner in which and for what purpose they are employed. Furthermore, different grades of blacktop and asphalt are available for use on different surfaces.

While there are numerous parallels between the two different types of paving, there are also significant differences. Understanding the distinctions will allow you to determine which surface is most appropriate for your project and will provide you with a working knowledge of these two widely used surfaces.

Similarities and Dissimilarities Between Blacktop and Asphalt
Allow us to first examine some of the parallels and contrasts between blacktop and asphalt before moving on to the distinctions.

The most striking similarity between the two products is that they are both created from the same two ingredients: crushed stone and bitumen. In addition to holding all of the crushed stones together, bitumen is responsible for the dark colour of blacktops and asphalt pavements. It is derived by the distillation of petroleum.

Bitumen is even employed in some roofing operations and projects, as well as in some other applications. Outside of blacktop and asphalt or other paving projects, it is most typically encountered in flexible roofing asphalt tiles, which are a type of flexible roofing asphalt tile.

However, when bitumen is mixed with crushed stone and heated above 250 degrees, the result is either blacktop or asphalt, depending on the ratios used and the temperature used. Both asphalt and blacktop surfaces are pre-mixed before being poured after they have been heated.

The ease with which asphalt and blacktop can be installed is a significant advantage that both of these surfaces share. In contrast to concrete and other porous surfaces that can take up to two weeks to dry, asphalt and blacktop are normally ready for use within two days after being laid.

The length of time required can vary depending on the weather conditions present during the paving procedure and in the days following. However, both blacktop and asphalt are a simple and effective approach to do the work in a short amount of time.

The only other commonalities are in their physical appearance and their intended uses. Both are dependable, universally applicable, and long-lasting for a wide range of paving operations.

Because of their dark colour, they are easier to maintain over the winter months as well. The sun easily heats the dark surface of the ice, causing it to melt more quickly. As a result, blacktop and asphalt surfaces are far safer than other types of surfaces such as concrete.

1. There are significant differences in the manufacturing processes of blacktop and asphalt.
When comparing and contrasting the differences between blacktop and asphalt, we will begin with the method by which they are constructed, which is the most significant distinction between them.

For this reason, it's easy to mix together blacktop and asphalt because they're both made from the same materials. However, there are a variety of variables that must be considered during the manufacturing process, such as temperature variations and mixing ratios.

In order to manufacture blacktop, significantly more stone is required than in the production of asphalt. Because it contains more stone, it requires a greater temperature than asphalt, typically around 300 degrees, which is significantly higher than the temperature required by asphalt. Asphalt is normally heated to a temperature of roughly 250 degrees.

What is the process of making blacktop?

When it comes to most blacktop combinations, natural stone becomes a vital component. Natural stone is responsible for the glistening appearance of blacktops in some areas. When you drive over a blacktop road or driveway, you will notice that there is almost a shine or sparkling appearance to the road or driveway.

The usage of natural stones, which are included into the crushed combination, gives the mixture a dazzling sheen.

The greater temperature at which blacktop is baked, as well as the differing stone ratios, contribute to its longer lifespan. Over time, blacktop will normally outlast asphalt in terms of overall durability. Due to the higher temperatures used in its manufacture, it has a more pliable surface that can be resealed rather than suffering sharp cracks and potholes as frequently as a concrete surface would.

What is the process of making asphalt?
Immediately following the refinement of fossil fuel, asphalt is produced directly from the heaviest constituents of petroleum. When dealing with such a thick and heavy substance, a cutting agent is required, which is where crushed stones come into play.

Because the crushed stones are employed as a cutting agent, the required consistency can be attained with minimal effort. The crushed stones and petroleum by-products are then combined in a huge drum to form a cohesive mixture. The drum is utilised to assist in maintaining temperature control for the mixing process, which must maintain a temperature of approximately 250 degrees.

The 250-degree mark is the bare minimum that must be met in the case of asphalt. However, the temperature must remain within that range rather than rising to greater levels. Temperatures ranging between 250 and 260 degrees are good, with greater temperatures having a negative impact on the finished product.

Asphalt has a long service life, but it does not have the same malleability as blacktop when it comes to bending.

2. Distinctions Between the Types of Blacktop and Asphalt Used For
In addition to their differences in appearance, the uses of blacktop and asphalt are also important distinctions. Initially, you might wonder: "How does the difference between a road, a driveway, and a playground manifest itself?"

Certain jobs will frequently favour one over the other, and each has its own set of applications and advantages.

Blacktop has a variety of applications.
The most common application for blacktop is for ordinary paving operations. Most people think of paving jobs as being the types of jobs that are normally associated with the term. This is also the reason why a huge paved surface is referred to as a "blacktop" by a great number of people. The term "blacktop" is used to describe any sort of asphalt pavement, regardless of whether or not it was really blacktopped.

The following are the most common uses for blacktop:

Driveways Roads Playgrounds
Basketball courts that are paved
Parking lots are available.
In parks or around neighbourhoods, there are paved pathways.
Many of these scenarios call for blacktop, which is preferred since it is easy to repair and, as a result, can last for a longer period of time as a result. When compared to some of the things that asphalt is used for, which we will discuss in further detail in the following section, they are all quite light weights in comparison.

Because these types of surfaces do not require as much weight capacity as other types of surfaces, they will naturally last longer. In addition, the pliable nature of blacktop makes repairs far more doable than with other materials.

In the short run, asphalt may prove to be a stronger and more durable surface than concrete. However, you might want to think about blacktopping as a "customer-friendly" surface. It is dependable and long-lasting, but it will also allow for more repairs over the course of its lifetime than asphalt does.

Blacktop can be bent and manipulated, and it is quite simple to repair. This is one of the reasons why it is so popular as a driveway option or as a simple pathway in a park or garden. No one wants to have to replace items of this nature on a regular basis. Because blacktop is more durable than asphalt, repair work will be less difficult.

In addition, the aesthetics of blacktop are frequently preferred for these areas. Using blacktop instead of asphalt to lay down a new basketball court in your backyard will look far better than using asphalt.

Asphalt has a variety of applications.
Asphalt is more widely recognised for its long-lasting properties and capacity to tolerate adverse weather conditions. Aside from that, it possesses water-resistant properties that are lacking in blacktops.

Asphalt is most commonly used for the following purposes:

Roadways, freeways, and highways that are major thoroughfares
Runways at airports
Coatings for cables
Soundproofing
Linings for swimming pools
Linings for reservoirs
Damp-proofing
It is easy to forget that asphalt is used for a variety of purposes other than paving in the traditional sense. When we think of roads, driveways, and other types of surfaces, we tend to think of asphalt or concrete. Asphalt, on the other hand, has additional use outside of the conventional realm.

One of the most significant contrasts between the two is that asphalt is utilised for non-paving projects, which is one of the most significant variances between them. Other than for paving operations, blacktop is not used for anything else. However, due to its greater versatility and extra characteristics such as water resistance, asphalt can be utilised for a wide range of other projects as well.

As a result of its water-resistant properties, it can be utilised for projects such as in-ground pools and damp-proofing. Furthermore, the substance's thickness makes it an excellent supplement to any soundproofing tasks you may be working on.

When it comes to roads, asphalt is normally reserved for heavy-duty roads that see a lot of traffic and carry a lot of weight. Despite the fact that blacktop is ideal for a driveway where you can park your two vehicles, it will fall short when it comes to parking a fleet of Boeing 737 aircraft.

Asphalt is widely used to construct highways with large traffic volumes and airport runways because of its exceptionally long life expectancy and durability. The greater the amount of weight that the surface will have to withstand, the more work asphalt will have to do.

3. There are several grades or types of asphalt.

When comparing the differences between blacktop and asphalt, it's important to remember that asphalt has a greater variety of textures and colours. While the appearance of blacktop may differ depending on the amount of natural stone used or the ratio in which it was mixed with other materials, there is only one type of blacktop.

However, when it comes to asphalt, there are several different grades to consider.

Asphalt Everlasting Pavement - The asphalt perpetual pavement form is a multilayered technique that is used in the production of asphalt. Because it is the most versatile method of installing asphalt, it can be compared to blacktop in terms of appearance and functionality. It is important to note that the foundation of the pavement is a flexible but incredibly robust layer that helps to avoid cracks in the pavement.

It is replaced on a regular basis, but it has a long lifespan as a result of the multilayer process and the additional flexibility. They can later replace simply the top if there is a problem with the surface level, which permits the surface to remain intact for a longer period of time before requiring a complete repair.

Quiet Asphalt - Quiet asphalt is exactly what it sounds like: it is quiet. Typically, this type of asphalt is found in residential neighbourhoods or on highways that are close to residential areas. They can lower road noise by as much as 50% by simply utilising a greater stone content mix of asphalt in the construction of the road. It is cost-effective, practical, and well-liked by homeowners who live in close proximity to a major thoroughfare.

Poured Asphalt - Poured asphalt is widely used for parking lots and other projects where it is desirable for water to be allowed to drain through to the earth beneath the surface of the asphalt. It is also utilised for storm water management in regions where there may not be adequate water management systems in place to allow roadways and other areas to drain appropriately.

Because of its durability and versatility, it may be used for a wide range of projects. However, it is most commonly employed in places that require water management and a porous surface that allows water to leak beneath the top.

Warm Mix Asphalt - Using a lower temperature for the heating of the asphalt throughout the production process, warm mix asphalt can help to save fuel expenditures during the production process. It helps to reduce greenhouse gas emissions while also allowing for the extension of the paving season.

Hot Mix Asphalt - Blacktop and hot mix asphalt are often confused with one another. The "hot mix" is what gives it its blacktop appearance. As a result, whenever you hear about someone using asphalt for a driveway or other similar surface, it is most likely hot mix asphalt, which is also referred to as blacktop in some circles.

These various grades of asphalt are vastly different from the usual hot mix/blacktop surface that most people are familiar with. They each provide a distinct function and can be put to a number of uses.

When it comes to asphalt grades, there are two different alternatives available to homeowners. You have the numbers 41A and 41B. They are similar in appearance, but they have distinct characteristics and functions.

41A can be made up of a mixture of rock and sand with a higher oil content of 6 percent than the other materials. The diameter of the rock and sand used is very small. This grade is commonly used for paving driveways and other related projects.

The rock and sand diameters in 41B are bigger, allowing this form of paving to resist greater weights than other types. When it comes to 41B, the oil content lowers to approximately 5%.

The smoothness of 41A and 41B is the most noticeable aesthetic difference between them. Because of the fine sand and rock in 41B, the surface will be significantly smoother and quieter than it was previously. Because of the larger elements in the composition, 41B tends to be a little bumpier, but it is also more durable.

When selecting which is the greatest option for you, the most important consideration will be the amount of weight load it will require. If you have a lot of needs, go for 41B. However, if you do not have a significant requirement for a large weight capacity, the smoother alternative of 41A will suffice.

Identifying the Differences Between Blacktop and Asphalt

Both blacktop and asphalt are safe, long-lasting, and extensively used paving techniques that may be applied to a wide range of projects and applications. Both blacktop and asphalt are durable, although blacktop is more flexible while asphalt is often the tougher of the two materials to work with. Their commonalities make them both some of the most commonly used paving materials, but their differences help you choose which will be the ideal pavement option for your project based on your needs and budget.

CHAPTER 17:- BUILDING MATERIAL :- CONCRETE GETS HOT

When was the last time you went for a barefoot walk on concrete under the blistering summer sun? It can get pretty hot in the summer. This is due to the fact that when sunlight shines on the ground, a significant quantity of energy is produced by the ground. As soon as our bare flesh comes into contact with the concrete, we are aware of the sensation.

The elements of concrete combine to form a thermal mass that absorbs the sunshine and heat that falls on it during the day. This generates energy, which is then released back into the atmosphere during the cool of the nighttime hours. If the concrete is exposed to direct sunshine or other types of heat, the concrete will become hot to the touch.

Around the world, concrete is used to pave much of the ground on which we walk or run on. What is the underlying cause of the temperature swings it experiences in response to the weather above it? It all has to do with the components used in the production of the concrete.

What Causes Concrete to Heat Up?

Concrete becomes hot as a result of the components that make it up: water, cement, and aggregate (sand, stones, and/or gravel). When sunlight shines on the surface, this mixture of materials undergoes a chemical reaction that results in the formation of a thermal mass, which absorbs the heat emitted by the sun.

The heat generated by concrete is also dependent on the stage of setting and curing at which the contents are placed.

Construction Materials Concrete is composed of the following constituents:

Portland cement water is used in the manufacturing of Portland cement (a general type of cement made up of materials such as limestone, sandstone, marl, shale, iron, clay & fly ash)
Sand, stones, and/or gravel are examples of aggregate.

In the process of making concrete, a chemical reaction takes place between the Portland cement and the water. This is the basis for the fact that concrete in its finished condition absorbs energy.

How Concrete is Constructed

Concrete must be allowed to cure and harden before it can be used in its best state of function, which we experience as our everyday walkways. When Portland cement and water are combined, as previously stated, a chemical reaction happens, which results in the formation of concrete. Water is added to the cement and the dissolving cement causes the release of hydroxyl ions from the dissolving cement.

Hydration is the term used to describe the reaction that occurs when water and cement come together. During the process of dissolving the cement, it raises the levels of calcium and silicon present in the solution. A precipitation reaction is subsequently initiated, which results in the formation of new solid products.

The amount of space that can be filled in between the cement and hydration products that have been produced determines the new strength that can be obtained by mixing the two materials together. It is possible that hydration will continue for several weeks or even years after the concrete has been hardened by this procedure.

The Effect of Concrete's Hydration on the Temperature of the Concrete

"Hydration is an exothermic process that creates heat through chemical reactions," says the National Institute of Health. HOW IT WORKS (from the source): Concrete Hydration generates heat as it continues to set and cure during the curing process.

It is a widespread assumption that the concrete is drying, however this is not the case. In reality, it is constantly undergoing chemical reactions that modify the water molecules, resulting in the formation of solids. The overall temperature of concrete rises as a result of this.

How to Lower the Temperature of Concrete

When it comes to heat regulation, different measures may be required depending on the form of the building or the type of structure that is being constructed with concrete. In the case of smaller constructions, such as sidewalks, the danger may not be as high because heat can dissipate through the soil or the air beneath them.

When dealing with considerably larger constructions, such as dams, some additional precautions must be taken. If the temperatures become too high, the structures may shatter because the heat generates internal expansion that is too great for the structure to withstand.

The maximum temperature differential between the interior and external concrete should not exceed 20 degrees Celsius (36 degrees Fahrenheit) while building such structures, according to the National Institute of Standards and Technology.

Why Extremely Hot Weather Can Create Issues for Cement

During the curing process of concrete, hot weather induces the evaporation of water. It is therefore required to add extra water to the mix in order to let the cement to react with the water and produce a concrete that is abundant, sturdy, and long-lasting.

The rate of temperature rise, the type of aggregate used, and the stability of the concrete can all have an impact on how well it performs at higher temperatures. As a result of thermal shock, sudden temperature changes in concrete can cause cracks to appear in the concrete. Distress can also manifest itself in the real environment.

When working on a concrete set, it is critical to keep track of the amount of water evaporating as a result of the heat and sunlight in the surroundings. If there is insufficient water, the cement will not adhere and the paste-aggregate link will be weakened.

Methods for Cooling Down Concrete for Use in Construction
In this chapter on the USDA website, it is suggested that construction workers reduce the temperature of concrete in order to "decrease water demand, slow slump loss, increase setting time, and reduce the chance of plastic shrinkage cracking" in order to "decrease water demand, slow slump loss, increase setting time, and reduce the chance of plastic shrinkage cracking":

Utilize fly ash and water reductions as cement replacements to keep the cement content to a bare minimum.
Retarders are used to control quick setting, and they can also be utilised in situations when extended haul durations cannot be avoided.
Instead of using water, third-generation high-range water reducers should be used.
Type II moderate heat cement should be used in place of Type I regular cement.

Shade aggregate stockpiles and moisten the cement to encourage evaporation and cooling of the mixture.
When working with extremely high temperatures, use cooled batch water.
Stay away from long truck waiting periods.
Use 100 drum revolutions during mixing, and make sure that the speeds established by the mixer manufacturer are appropriate.
Delay the mixing process until the truck arrives at the jobsite (primarily for long hauls in hot, humid weather)
Paint the mixer drums white to prevent them from absorbing too much heat.
Deliveries should be scheduled to avoid the warmest parts of the day. Extra water should not be used on the job site.

Concrete Structures and Design Concepts
Construction of concrete buildings and structures can be found all over the world, in a variety of various styles and designs. Some of these are as follows:

Walkways
Bridges
Towers
Dams
Some examples of ancient architecture
The constructions described above are just a few examples of the numerous varieties of concrete structures that exist. As previously said, concrete becomes heated not only during the setting and curing processes, but also as a result of the heat that has accumulated and been absorbed from the surrounding weather conditions and direct sunlight.

Concrete shrinkage and cracking are two common problems.

Consider this the next time you're walking along a concrete street: it's worth taking a closer look at it. You might see a number of cracks, some of which are larger and some of which are smaller.

This is a very typical occurrence and should not be taken seriously. When it comes to larger constructions, however, it is necessary to provide more constant care and maintenance to ensure that the concrete is intact and strong.

Shrinkage
"Drying shrinkage is defined as the contraction of a hardened concrete mixture as a result of the loss of capillary water," according to the American Concrete Institute. (Image courtesy of Drying Shrinkage) As the concrete slab continues to cure over time, it loses moisture in the process, resulting in the slab becoming smaller in size.

In the case of typical concrete walkways, the shrinkage is not particularly important, especially if the cracks that arise from the shrinkage are minor or medium in size.

With bigger concrete constructions or buildings, however, it is important to consider what other structures are supporting the concrete and how they are holding the concrete in place (such as stabilising joints, for example).

Buildings that shrink require regular care to keep them in good condition. You may find an example of how to preserve the concrete of masonry walls here, which goes into greater detail.

Cracks

Cracks in concrete are frequently caused by the addition of too much water to the mix. The primary reason for increasing the amount of water used is to make home work a little bit easier to complete. Having said that, the addition of water does not inevitably result in a construction that is more stable and devoid of cracks.

The most common cause of cracking is shrinkage, and when shrinkage happens, the slabs of concrete are forced apart, resulting in the formation of cracks.

So, what causes concrete to become hot?
What does concrete cracking and shrinkage have to do with the temperature of the concrete you may be wondering. They are, in fact, very close in relationship. Overheating of the concrete results in fast water evaporation, which leads to shrinkage and eventually breaking.

The upkeep of concrete structures may necessitate more frequent care in places with increased heat and humidity as a result of this. You should now be aware of what is going on beneath your feet while you stroll to the park or run to the school bus stop.

The heat generated on the ground is not just a result of the sun's rays, but also of chemical processes occurring within the concrete itself as hydration takes hold, resulting in a stronger structure.

CHAPTER 18:- BUILDING MATERIALS :- ADMIXTURE

In addition to Portland cement, aggregates, and water, admixtures are substances that are added to concrete to improve its strength and durability. Despite the fact that they are not essential, they are occasionally added to improve the concrete mix on a selective basis.

Admixtures are used in concrete to change the properties of the material in a variety of ways. Some of the most typical applications include improving workability, increasing or decreasing cure time, and raising the strength of concrete. It is also possible to utilise additives for purely aesthetic purposes, such as to change the colour of the cement.

Selecting appropriate types of aggregate and Portland cement and keeping a proper water-cement ratio should be used to provide the best possible workability, water tightness, and strength in concrete wherever possible.

Admixtures can be beneficial when this is not possible or when unusual circumstances exist, such as freezing weather, high temperatures, increased wear, or prolonged contact to deicing salts or other chemicals.

Admixtures in Concrete Come in a Variety of Forms

The sorts of admixtures that are currently accessible vary widely in terms of their functionality and intended use. The following is a list of some of the more frequent types:

Admixtures that encourage the movement of air
Admixtures that reduce the amount of water used
Superplasticizers
Admixtures that accelerate the process
Admixtures for Retarding
Silica from Fly Ash Fume\sBlast Furnace Slag is a type of slag produced by a furnace.
Pozzolans
Agents of Workability
Corrosion Inhibitors are substances that prevent corrosion from occurring.
Admixtures for Bonding
Hardeners
Coloring Agents are substances that give colour to things.
Agents that cause gas formation
Additions can be made either before or during the mixing process. Because they alter the physical qualities of concrete, they should only be used in limited quantities and only under the supervision of a concrete professional.

Admixtures that encourage the movement of air

It is necessary to apply air-entraining admixtures in order to entice small air bubbles into concrete. This increases the durability of the concrete by improving its resistance to damage caused by freeze-thaw cycles as well as from deicers, which induce scaling, as well as from other environmental factors.

In addition, air entrainment improves workability during insertion and has greater water tightness compared to other materials. As a bonus, it uses less water per cubic yard of concrete than non-air-entrained concrete, resulting in an improvement in the water-cement ratio.

Air-entraining admixtures contain a variety of active components, including polyethylene oxide polymers, certain fats and oils, sulfonated compounds, and detergents, among others. ASTM C226 is the standard that specifies air-entraining admixtures.

Admixtures that reduce the amount of water used
Concrete admixtures that lower the quantity of water required to manufacture concrete of a specific consistency are known as water-reducing admixtures. Besides that, they can be utilised to increase the slump of concrete without the need for additional water, which results in a lower water-to-cement ratio.

When the water-cement ratio is reduced, water-reducing admixtures can give increased concrete strength while maintaining the same workability as the original mix design. Water-reducing admixtures such as lignin sulfonic acids and metallic salts are widely employed in the construction industry.

Superplasticizers
Inorganic additives known as superplasticizers turn a stiff concrete mix into one that flows more easily. They function by coating cement pchapters, helping them to break away from the lumps that are generally formed when cement and water are mixed together.

As a result of the superplasticizers, each cement pchapter has a negative charge, which causes them to reject one another, resulting in a more even dispersion of the cement. The purpose of these devices is to either make the placing of concrete easier or to reduce the water content of a concrete mix in order to increase the strength of the concrete.

Admixtures that accelerate the process
Accelerating admixtures, often known as accelerators, are used to speed up the curing of concrete. Additionally, they help to speed the development of concrete's strength. The use of Type III high-early-strength Portland cement can help to expedite the development of strength even more.

Additional acceleration can be obtained by curing concrete at higher temperatures or by reducing the water-to-cement ratio during the curing process. Accelerators are often utilised in cold weather to help the body gain strength more quickly in order to combat the possibility of freeze damage.

Admixtures for Retarding
Accelerating admixtures are essentially the polar opposite of retarding admixtures. These additives lengthen the time required for concrete to cure, or in other words, they reduce the amount of time required for the cement paste to set. In hot temperatures, heat-retarding admixtures are frequently employed.

It is possible to reduce the amount of water necessary to cure the concrete mix by employing retarders in hot weather. In the absence of a retarder, more water would be required to achieve the necessary slump, resulting in lesser strength concrete.

Using retarding admixtures, it is possible to create a superior water-to-cement ratio as well as higher final strength.

Fine powdered debris produced by coal-fired power stations, fly ash is a waste product that must be disposed of properly. This additive can be used to boost concrete strength, decrease permeability, and increase sulphate resistance in many applications.

Other advantages of fly ash include the fact that it can assist in lowering the temperature rise during curing and lowering the amount of mixing water required. Furthermore, it has the potential to increase the pumpability and workability of the concrete.

Silica Fume is a type of gas that is produced by the breakdown of silica.
Silica fume, also known as microsilica, is a powder that is produced during the production of electrical semiconductor chips. It has a finer texture than Portland cement and is predominantly composed of silicon dioxide, as opposed to Portland cement.

In the presence of cement, it can form concrete with extremely high strength and low permeability when mixed with other ingredients in a concrete mix. As a result, it is an excellent choice for places with high moisture content or regions with high water tables.

Blast Furnace Slag is a type of slag produced by blast furnaces.

Blast furnace slag is a byproduct of the iron-making process that is used to make steel. It can be used as an aggregate to improve the workability and strength of concrete, as well as to increase its durability.

Additional advantages of blast furnace slag include lower permeability, a reduced rise in temperature during curing, and improved sulphate resistance, amongst other characteristics.

Pozzolans
Pozzolans are a siliceous or a combination of aluminous and siliceous materials that, when finely powdered and coupled with water, may react chemically with calcium hydroxide to generate compounds with cementitious qualities. Pozzolans are commonly used in the construction industry.

The use of pozzolans in a concrete mix has several advantages, including lower costs since less Portland cement is used, reduced environmental pollution because less Portland cement is produced, and greater durability.

Agents of Workability
Working agents, as their name implies, are substances that are employed to improve the workability of a concrete mixture. The usage of this method is especially beneficial in circumstances when new concrete is harsh due to faulty aggregate grading or wrong mix proportions.

Where a troweled finish is required on concrete, workability is critical, and the use of workability agents can be advantageous in these cases.

Corrosion Inhibitors are substances that prevent corrosion from occurring.

Corrosion inhibitors are typically used in concrete mixes to reduce the rusting of reinforcement steel, which is a good thing. When it comes to structures that are in close proximity to highways or that are a component of the road infrastructure, this can be very valuable.

In the winter, when deicing salts or other corrosion-causing chemicals are placed on highways to keep them clear of ice, they can hasten the rusting of reinforcement steel in buildings. When corrosion inhibitors are put into concrete structures under these conditions, they can be quite advantageous.

Admixtures for Bonding
Occasionally, it is necessary to pour new concrete over existing concrete in order to repair or replace it. Because the existing concrete has already reached its full setting point, it is difficult to build a firm link between the new and existing concrete.

In order to improve the link between freshly placed and fully set concrete, bonding admixtures are employed. It is normally included into the new Portland cement mix, but it can also be applied on top of the old concrete to give it a more polished appearance.

Hardeners

It is necessary to apply Hardening Admixtures in instances when a concrete surface is going to be subjected to a great deal of wear and tear. A manufacturing or warehouse floor, for instance, could serve as an illustration. The life of the floor can be prolonged by applying a hardener, which can be either a liquid or a dry-powder formulation.

When used with Portland cement, liquid hardeners operate by causing a chemical reaction with the free lime and calcium carbonates in the cement. Dry-powder hardeners can be applied to freshly poured concrete using a dry shake technique.

Colorants and Coloring Admixtures

Colored concrete is frequently specified in architectural drawings and specifications. Coloring admixtures can be used to add colour to a concrete mix in a variety of hues, albeit they are often limited to earth tones and pastel shades.

A pure, finely powdered mineral oxide can be used to colour concrete by mixing it with Portland cement in the proper proportions. When new ingredients are added to the mix, thorough mixing is essential in order to achieve a consistent appearance. It is also possible that the mixing time will be longer than usual.

Agents that cause gas formation

Concrete can be treated with gas-forming agents to allow for a modest expansion of the material before it hardens. Aluminum powder is a typical gas-forming additive that is utilised in a variety of applications.

Small hydrogen gas bubbles are produced as a result of the reaction with the hydroxides in hydrating cement. This also aids in the elimination of any voids that may have formed as a result of concrete settlement.

Concrete Admixture Content as a Percentage

The percentage of admixture used in concrete varies based on the type of additive used and the amount of cement used in the concrete. Typically, the dosage is specified by the manufacturer and is determined by the amount of cement present.

Overall, admixtures are included in tiny proportions in concrete mixes, with values of 5 percent or less being the most commonly encountered. When possible, it is recommended to make a test batch of concrete before mixing the concrete, or to build mockups to identify the most effective mix ratio.

Admixtures in Concrete Have Several Disadvantages

While admixtures have numerous advantages, as has been discussed thus far, there are some disadvantages to employing them as well. For starters, they have a tendency to raise prices. They also have the additional effect of altering the composition of concrete, and as a result, should be utilised with caution.

Admixtures are not always straightforward to work with, at least not until you become more familiar with how they are to be utilised. As a result, it's usually a good idea to collaborate with a concrete professional who is well-versed in the material's application.

In summary, additives are employed in concrete to improve the overall performance of the mix in a variety of different ways. Admixtures, which are typically added before or during the mixing process, can have a variety of beneficial effects, including increasing the strength of the mix and speeding up or slowing down the curing process, among others.

CHAPTER 19:- BUILDING MATERIALS :- STONE (MARBLE, GRANITE AND MORE..)

Because of their toughness and longevity, stones have been employed as foundational elements in construction for thousands of years. While some stones, such as granite and basalt, are prized for their greater strength and durability, others, such as gneiss, are frequently utilised for their aesthetic appeal and ornamental qualities. With a variety of stones to pick from, it's critical to understand which one would work best for your project.

There are many different kinds of stones that are utilised in construction. Basalt, granite, and sandstone are high-compressive-strength stones that are appropriate for large-scale construction projects such as dams and bridge piers because of their high density. Stones such as travertine, gneiss, quartzite, and marble are excellent choices for modest construction and décor projects.

While some are ideal for use in the construction of walls and foundations, others are better suited for use as ornamental elements. Continue reading as we take a look at some of the most commonly used stones in construction. Furthermore, in order to provide additional information, the essay will also emphasise the important requirements for good construction stone.

1. Basalt Stone (also known as basalt rock)
Basalt stones, also known as traps, are igneous rocks that form when molten lava hardens under pressure, resulting in the formation of a trap. It is because of their strength and compactness that basalts are used for major construction projects and as aesthetic elements in homes.

Basalts have a delicate texture that, at times, reveals pores and cavities caused by the escape of gases from the lava as it cools throughout the cooling process. Because of the concentration of ferromagnesian minerals in basalt, the majority of basalts have a light-dark or dark look in appearance.

Granular basalt has a compressive strength that ranges from 200 MPa to 350 MPa, making it a strong and compact material that is ideal for major construction projects. Basalt is extremely resistant to water, wind, and other components of the environment. Basalt, on the other hand, has a strong compressive strength, which makes it difficult to shape or dress into small slabs.

Granite is the second most common type of granite.

Other than the fact that it is extremely durable, granite has a compressive strength that ranges between 100 MPa and 250 MPa. Granit's strength and toughness are enhanced by the inclusion of quartz, feldspar, and mica, which makes it an excellent choice for both modest and large-scale construction projects. Granite can be used in heavy-duty building projects like as retaining walls, stone pillars, curbs, railway patience, exterior wall cladding, coarse aggregate in concrete, and dams, among other things.

Granite stones are well-known for having low porosity, a low absorption value, and excellent weathering and weathering resistance. Despite the fact that graphite has a low fire resistance rating, it can be polished to a glossy, fine finish, making it an excellent material for use in interior décor. Granite will also last for a long time if it is used in acidic areas since it contains quartz and feldspar, both of which are resistant to the effects of acid rain.

3. Sandstones are a type of rock.

Sandstone is composed primarily of quartz pchapters that have been eroded from nearby rocks (typically granite) over a prolonged period of time. After that, the quartz pchapters combine with minerals contained in natural cement to form a solid mass.

Given the presence of accessory minerals such as feldspar, micas, and sometimes even dark minerals, you can find sandstone in a variety of colours ranging from white and light grey to brown and dark grey. You can also find it in a variety of colours including grey, white, yellow, brown and red and even buff. Specific gravity for sandstone is 1.85 to 2.7 and compressive strength is 20 MPa to 170 MPa, with the range being 20–270 MPa.

Sandstones are frequently utilised in the construction of big structures, particularly in conjunction with high-quality silica cement. Sandstones are also used in the construction of dams, river barriers, bridge piers, and a variety of other masonry projects.

4. Slate

Fine-grained metamorphic rocks such as slate are generated when sedimentary rocks are subjected to high pressure, and slate is one type of metamorphic rock. One of the most advantageous characteristics of slate is its capacity to be divided into sheets with relative ease, which justifies its use in the majority of roof coverings.

Due to its adaptability, slate may be used for a variety of applications other than roof coverings. Pavements and slabs are two examples. The granite is far too strong and is considered to be one of the greatest stones available for protecting against possibly dangerous weather factors.

Slate is unique in that it may be used for a variety of applications while not being as strong as graphite when carrying heavy loads. The stone may be divided into appropriate-sized slabs along its foliated planes, allowing it to be used in a variety of applications such as roofing, flooring, damp proof courses, and partitions, among others, because of its versatility.

5. Limestone

It is possible to find several varieties of limestones, some of which are not suitable for use in building construction. The ideal limestone for use in construction should be compact, dense, fine-textured, and devoid of cracks and cavities, and it should also be free of fractures and cavities. It is frequently the case that undesirable limestone kinds are soft and high in clay content, making them unsuitable for use in construction. Limestone is known to include calcium carbonate as one of its constituents. It also contains calcite, which is formed over time by the decomposition of marine creatures. The presence of calcite indicates that limestone has fossils as one of its constituents, which makes it easy to polish or hone.

Calcium, a raw material for cement, is also used to construct ceilings, floors, sidewalks, cladding for buildings, bathroom wall tiles, and vanity tops. It is also utilised to manufacture concrete. Building architects and structural engineers are also interested in incorporating limestone into the fabric of structures for use as pillars, cornices, facades, and a variety of other aesthetic elements.

A disclaimer: Because of its sensitivity to pollution from industrial gases, limestone is not recommended for use as a face stone. Because of the frequent attacks from salty winds, limestone used as a face stone in coastal areas can develop a rustic, ugly appearance over time.

6. Laterite

Laterites are sedimentary rocks that are primarily composed of aluminium oxides, with various levels of iron oxides present as well. Known to occur as a result of igneous rocks' chemical disintegration (alkaline) due to the leaching of some basic components, the rock has a spongy or porous texture and is typically found in sandstones and limestones.

The colour of laterites is usually determined by the amount of iron present, with the majority of laterite stones having brown, yellow, red, or grey hues. Laterites are known to develop stronger over time, despite the fact that they are not the strongest of stones.

While laterite can be used as a building block in construction, it must be coated on the outside in order for the stones to withstand the weight of the structure. The stone is available in both hard and soft variants, and it is frequently utilised in pavement construction as well as a variety of masonry projects.

7. Marble

Marble is one of the most widely used ornamental rocks in the world, and it is often distinguished by its crystal-like characteristics. Generally speaking, the stone is formed primarily of dolomite or calcite crystals, and it typically takes on a crystal-like look as a result of the combination of pressure and heat over time. Its widespread use in construction projects can be attributed to the fact that marble is a metamorphic rock that is both compact and robust. The stone is available in a variety of colours, with the majority of them ranging from deep black to pure white. Marble's colour, like that of other stones, is highly influenced by the presence of minerals and impurities in the stone, particularly during its production.

A marble's low porosity, homogeneous texture and ability to be polished make it a popular choice for flooring and other surfaces. Marble is a beautiful ornamental stone because of its unique characteristics. Marble may be found in a variety of applications, including bathrooms, floor and wall tiles, work surfaces, shower trays, cladding, and even staircases.

Marble can also be used to create work surfaces for the kitchen. The downside to marble is that material is typically more vulnerable to stains, wear, and tear than granite.

8. Gneiss

In addition to beautiful stones, gneiss is often used in flooring, gravestones, and the facing stones of constructions. However, due to the fact that the stone comes in a variety of robust variations, it can also be utilised for light construction work.

In addition to purple, grey, pink, dark grey, and greenish-gray, gneiss can be found in a variety of other hues, which are determined by the minerals or impurities present at the time of formation. It is possible to utilise gneiss stones as dimension stones if they are strong and durable enough. After they have been sawn or seared into slabs and blocks, they can be used in a wide range of curbing, construction, and paving operations.

Crushed gneiss, despite the fact that it lacks the compactness and hardness of granite, can be utilised in light building projects, stone pitching, and even rough stone masonry due to its low density. Despite the fact that it is not the most popular stone on the market, it may give structures a unique edge and completely change their appearance.

9. Quartzite

Quartzite is more stain and scratch resistant than granite, which is why it is frequently used for kitchen countertops rather than granite or other natural stone. Quartzite is also an excellent material for floor tiles, walls, and stair treads. The stone is essential in the manufacturing of industrial silica sand, and it is also an excellent choice for railway ballast.

Quartzite has a fine to coarse grain and is primarily branded and granular in appearance, with some crystalline inclusions. Yellow, grey, and white quartzite are just a few of the colours that may be found in this mineral. Although the stone is not recommended for heavy construction, it is well-known for being a good (although expensive) option for interior decorating projects.

10.Travertine

Travertine is characterised by the presence of troughs and pitted pores on its surface, which makes it extremely porous. However, because of the stone's internal qualities, it is easily polished, which is ideal when used as an ornamental accent for interior decorating projects such as furniture.

The stone is a type of limestone, and it is generated as a result of the quick precipitation of calcium carbonate. Despite the fact that travertine cannot be used for heavy construction, it may be used for a variety of light projects like flooring, wall cladding and facades, shower trays, vanity tops, stands, and basins, among others.

11.Alabaster
Alabaster has been used in construction for numerous centuries, with the stone being particularly popular throughout the mediaeval period for altars, effigies, tombs, and religious carvings. Despite this, because of its translucence, the stone is currently extensively utilised for light to medium-sized fixtures.

Due to the nature of alabaster, it is not weatherproof and will deteriorate significantly if subjected to factors such as severe rainfall, strong winds, and snow over an extended period of time. Because of its translucent nature, some individuals even use alabaster to construct windows.

Therefore, because of its suppleness, alabaster may be easily shaped into many shapes. Please keep in mind that alabaster is extremely sensitive to moisture, and that extended contact to moisture will result in the stone being duller, rougher, and less translucent than typical.

The Most Important Requirements for Quality Construction Stones

Not all stones are suitable for use in big construction projects. A similar set of materials will be required for interior (or exterior) decorating and finishing, which will include aesthetically pleasing stones that can be easily transformed into a variety of shapes and patterns.

Let's have a look at some of the most important factors to consider while buying construction stones.

The Structural Strength of the Building Stone
The stones used in construction should be sturdy enough to withstand both static and dynamic loads. When working on large-scale construction projects, it is strongly recommended that you employ stones with high compressive strength as building materials. The use of stones with a lower load capacity indicates that the structure will not be strong enough to support the total weight of the structure, raising issues about structural integrity.

In order to be suitable for large projects (such as foundations for houses), a construction stone's optimal strength should be in the range of 60-200N/mm2. Extra care should be taken in selecting a stone that is both strong enough to support the weight of your intended building and compact enough to sustain years of dead and dynamic loads.

Durability
Weathering stones are constantly subjected to the elements such as rain, wind, and sunlight, as well as vibrations caused by ground movements. As a result, you should choose stones that are robust and durable enough to withstand the harmful impacts of the aforementioned chemicals and other environmental factors.

The chemical makeup of construction stones, as well as the chemical elements present in the immediate surrounding atmosphere, are important factors in determining their long-term durability. Additionally, the texture of a stone is important in determining the durability of that particular stone.

In order to be on the safe side, use stones with dense structures that are close-grained and crystalline uniform in composition. These stones can withstand considerable weight and are sturdy enough to withstand progressive deterioration caused by wind, rain, and other environmental factors.

Hardness and toughness are two characteristics of a material.

The hardness of stones should be considered in addition to their durability, particularly if they are intended for usage in high-traffic environments. Flooring, bridge aprons, and roads should all be made of strong, durable stones that can stand up to friction and progressive wear without cracking or crumbling.

It is recommended that you utilise the Mohr's scale of hardness to determine the toughness and hardness of the stones that you intend to use in your construction project. You can also use a scraping motion to determine the hardness of the stone's surface. After extensive scraping, hard stones almost never show any signs of wear.

Specific Gravity is a measure of how heavy something is. In general, the higher the specific gravity of a stone, the stronger and more capable the stone is of withstanding massive weights, as a rule of thumb. As a result, while working on heavy-duty construction projects such as dams, docks, ports, and retaining walls, specific gravity-rich stones are the best choice.

Possibilities for Porosity and Absorption
Did you know that the porosity of stones is usually determined by the structural structure of the parent rock as well as the mineral components present? As a result, it is critical for building projects (particularly those involving exposed stones) to choose non-porous stone because it is typically harder and more resistant to progressive disintegration than porous stone.

Generally speaking, stones with high porosity ratings absorb more water than their less porous counterparts. The use of solid, impervious boulders in heavy-duty construction projects illustrates why this is a necessary practise. In cold climates, rainwater frequently freezes inside the pores of the stones, causing them to disintegrate as the stones grow more porous.

As a result, conducting porosity tests to identify how porous a stone is before utilising it in building is highly recommended. Please keep in mind that porous stones should only be used in strategic positions in building, particularly in areas where they are unlikely to be subjected to rain, frost, or excessive moisture during the construction process.

Seasoning and dressing are two important aspects of cooking.
The best construction stones should be well-seasoned and free of quarry sap when they are harvested from the quarry. Quarry sap can be eliminated by placing the stones in a shaded area with no walls, which allows for slow drying and air circulation to take place. Disclaimer: Seasoning times vary depending on the type of rock used, with lateritic stones requiring at least several months to season following quarrying.

Before employing a stone in a construction project, it is important to determine how well it can be dressed in the process. The stones should be easy to cut into blocks and square in order to meet the requirements of the specifications. To season your stones, you'll need instruments such as a hammer, puncheon, axe, and chisel, among others. The expense of dressing stones during construction can be extremely expensive, which is why it is important to use stones that are simple to season when building.

Appearance
The appearance of stones should always be checked before making a purchase decision. Ideally, the hue of the stones should compliment the colour of your main design element. Furthermore, as most architects will agree, when dwellings and commercial buildings are designed to blend in with their surroundings, they generally appear better.

The use of lighter colours rather than darker ones when selecting colours is highly suggested when designing a room. This is due to the fact that brighter hues are typically less vulnerable to fading as time goes on. The strength of stones, on the other hand, should not be determined just on their appearance, which means you'll need to look for these characteristics independently. When it comes to stones used for face work and other decorative uses, appearance is more important than functionality. Before selecting a set of decorative stones for your interior or exterior décor, consider the colour coordination, attractiveness, and quality of the shades to guarantee that you obtain a long-lasting product that will last for years.

Fire-Resistant Materials

Because fire-resistant stones increase the ability of a building or structure to withstand fire and other harsh circumstances, they increase the value of the building or structure. The use of fire-resistant stones in the construction of a home can also assist homeowners obtain lower home insurance premiums.

As a result, when shopping for fire-resistant stones, make sure to look for products that are free of iron oxides, calcium carbonate, and minerals with varying thermal expansion coefficients when possible.

Cost
The cost of stones is a significant element that should be taken into account well in advance of construction beginning. Some stones are more expensive than others because of their versatility, strength, and durability, while others are less expensive because of their lower cost. The overall distance between the quarry and the customer is also important in calculating the total cost of the stones. A quarry's proximity to the building or structure being constructed reduces the total cost of construction since transportation costs are decreased..

Structure
The alignment of a stone's constituent minerals is indicated by the texture of the stone. As a result, a good building stone should have a consistent structure throughout. Due to the fact that stones with crystalline and uniform textures are typically compact and hard in comparison to their open-texture, non-crystalline cousins, this is the case.

When it comes to stones intended for use in major construction projects, they should be devoid of defects such as cracks, patches, cavities, and loose material. It is important to use fine-grained stones for ornamental carvings if you want to attain the desired aesthetics.

Weathering
A good stone should not weather even after being exposed to the elements for an extended period of time, including rain, sunlight, wind, and even snow. And, in order to establish the weathering capabilities of a particular stone, it is strongly advised that you examine historical buildings made of similar materials as well. You should investigate structures in an environment that is similar to the one in which you intend to construct.

If the sharp angles and edges of the historic buildings you study have preserved their sharpness, it is likely that the stone will continue to perform well despite lengthy exposure to rain, intense sunlight, and high winds.

Weight
When working on building projects, you may also need to take the weight of the stones into consideration. Heavy stones are known to have high specific gravities, in addition to being compact and less porous than lighter variations of the same type.

Docks, barrages, dams, and retaining walls are all examples of heavy-duty construction that benefit from heavy-weight stones. Lighter forms of stone, on the other hand, tend to work well for light structures, such as domes and roof coverings. As a result, the weight of the stones you employ should be dictated by the nature of the work being undertaken.

Rocks and stones are classified according to their composition.
There are three basic types of classification for rocks: geological classifications, physical classifications, and chemical classifications. Typically, the classification of rocks is essential in determining the qualities of a particular stone's characteristics. The following is a basic overview of the most common rock classifications.

Classification of Geological Structures
Geological classification of stones is mainly determined by the physical changes that have occurred, such as melting and cooling, progressive erosion, compacting, and finally deformation. According to geological classification, the following are the three categories of rocks:

Sedimentary Rocks are rocks that have formed through time.
Most sedimentary rocks are composed of fragments or sedimentary deposits of other rocks or biological substances. As nearby stones continue to erode throughout time, the bits or sediments that form are formed. Sediment transport is accomplished by the use of meteorological agents such as wind, gravitation, frost, and water.

Sedimentary deposits are formed as a result of the progressive accumulation of sediments, which result in the formation of layered structures. Clastic sedimentary rocks are the most common type of sedimentary rock, followed by organic sedimentary rocks and chemical sedimentary rocks. Organic sedimentary rocks are formed from biological components such as shells, plants, and bones, with coal serving as an excellent example. Clastic sedimentary rocks, on the other hand, are formed when fragments or clasts of other rocks weather through time, resulting in the formation of limestone. Chemical sedimentary rocks are typically formed as a result of chemical precipitation, and they frequently contain stones such as halite, flint, and limestone.

Igneous Rocks are rocks formed by volcanic activity. Igneous rocks are formed when magma solidifies beneath the surface of the earth. When lava fails to erupt, it cools and solidifies beneath the surface, and this is when igneous rocks are formed. Normally, the strength of igneous rocks varies depending on the depth or position of the magma at the moment of solidification. This is known as the depth effect.

When magma hardens deep beneath the earth's surface, the resulting rocks are referred to as plutonic rocks, and they typically have a coarsely grained crystalline structure, such as granite, which is an example of this.

In a similar vein, magma that hardens at an average depth generates hypabyssal rocks, which have finely grained crystalline structures and are composed primarily of quartz. Finally, volcanic rocks are rocks that originate close to the earth's surface and are often fine-grained in texture and composition.

Metamorphic Rocks are rocks that have undergone transformation.
Metamorphic rocks, as the name implies, are formed as a result of slow metamorphism occurring over time. Metamorphism is a natural process that occurs when stones change their features as a result of pressure and high temperatures. It is possible that the pre-existing rocks are either sedimentary or igneous in nature. Metamorphic rocks include gneiss, slate, marble, soapstone, and schist, to name a few of the most frequent types.

Physical Classification is a term that refers to the classification of objects in space.
According to physical classification, there are three major types of rocks: stratified rocks, unstratified rocks, and foliated rocks. The following is a quick description of the rocks mentioned above.

Rocks that have been stratified
Stratified rocks are distinguished by the presence of distinct strata that are divided by cleavage planes. The cleavage planes, also known as bedding planes, are the weak points in the rocks that allow them to be readily fractured and broken apart from one another. Limestone, shale, and sandstone are examples of sedimentary rocks that are commonly found.

Rocks that are not stratified

Unstratified rocks are more likely to have a crystalline structure than stratified rocks. These rocks are typically characterised by a consistent structural pattern over their whole body. If you look closely, you'll notice that most igneous and sedimentary rocks can be classed as unstratified rocks. The best examples are granite, trap, and marble, among others.

Rocks that have been foliated
Foliated rocks are characterised by a banded or layered structure that develops as a result of extended exposure to high temperatures and pressure. Unlike stratified rocks, foliated rocks can only be divided in a single direction, as opposed to stratified rocks. Foliated rocks include metamorphic rocks such as gneiss, slate, and schist, which are metamorphic rocks that have been metamorphosed.

Chemical Classification is a term that refers to the classification of chemicals.
Rocks can also be classified based on the chemical composition of their constituents. According to the chemical composition of rocks, there are three primary types: argillaceous rocks, calcareous rocks, and siliceous rocks, which are addressed more below.

Argillaceous Species of Rock
Argillaceous rocks are characterised by the presence of clay as the major component, which makes the resulting stones softer and more prone to crumbling. Laterites, slate, and shale are all examples of argillaceous rocks that are commonly found.

Calcareous Rocks are a type of rock that contains calcium carbonate.

Calcareous rocks are composed primarily of calcium carbonate, which is a naturally occurring mineral. Although calcareous rocks are often durable, the durability of calcareous rocks is often dependent on the composition of the rocks and plants in their immediate vicinity. Calcareous rocks include limestone, dolomite, and marble, to name a few examples.

Siliceous Rocks are a type of rock that is made of silica. Siliceous rocks are composed primarily of silica, which is the most abundant element. Because of the presence of free silica (in abundance), siliceous rocks are strong and robust, and they are resistant to weathering throughout time. Quartzite, chert, and granite are all examples of siliceous rocks that are commonly found.

Conclusion
As we've seen in this reading, there are many different sorts of stones that can be used in construction, each with its own set of features that make it unique.

The best stone for use in construction should be one that is appropriate for the project at hand. For major projects, for example, you'll need hefty stones with a high specific gravity and a low porosity to ensure a solid foundation. When it comes to beautifying or decorating your construction, ornamental rocks such as marble are frequently used.

Therefore, as you choose construction stones, it's best advised to consider some of the key factors such as seasoning, dressing, appearance, strength, durability, porosity, and specific gravity.

CHAPTER 20:- BUILDING MATERIALS :- M SAND

Sand of high quality that is workable and durable is essential in all construction projects, including the construction of residences, public buildings, and even specific roadways. Concrete can be found almost everywhere, and construction businesses require sand to produce it. However, as a result of a global scarcity of natural river sand, firms have begun to manufacture sand as a substitute. Is this, however, beneficial for construction?

M-sand is beneficial in the building industry because it produces concrete that is durable, workable, and high in strength. Its drawbacks include a coarse texture and a high concentration of micro-fine pchapters in the mix. M-sand, on the other hand, is environmentally friendly, cost-effective, and efficient, which is why the majority of construction companies are now utilising it.

Consider employing M-sand for your next building project if you haven't done so already. So, let's go into the features, advantages and cons of this particular substance!

What Exactly Is M-Sand?

M-sand, often known as manufactured sand, is a substitute for natural river sand in some applications. While river sand used to be the most commonly used resource in the production of concrete, a global scarcity of this material has forced construction businesses to look for new sources of materials.

M-sand is one of the most durable, cost-effective, and environmentally friendly of the several materials that are currently available on the market today. While M-sand has a texture that differs from natural river sand, it has several advantages over natural river sand that have made it the preferred alternative in the building industry.

In the manufacturing of hard granite rocks, quarry stones, and other aggregates, M-sand is produced by crushing the rocks into a sand-like substance. It is necessary to crush and blast M-sand numerous times during the manufacturing process in order to acquire a consistency that is close to that of natural river sand.

The sand that is produced as a result of this procedure is sieved to remove any remaining contaminants before being washed several times. Because M-sand does not naturally retain moisture in the same way that river sand does, the washing process must be used to add the moisture required for concrete blending.

Once completed, the final product is subjected to a series of audits to confirm that it is capable of producing high-strength, long-lasting concrete.

What Is the Purpose of M-Sand?

M-sand is believed to be a legitimate, more cost-effective, and more environmentally friendly substitute for natural river sand. As a result of the growing global scarcity of natural sand, it has become increasingly important in the construction industry.

In part because M-sand is readily available in most regions and because the expenditures of transporting sand from a suitable river bed can be avoided, a growing number of construction businesses are beginning to favour it over conventional natural sand.

What is the benefit of using M-Sand in construction?

There are a variety of factors that contribute to this product's widespread use in the construction sector and beyond. Here are a few examples.

It contributes to the reduction of natural sand scarcity. The building industry must continue and flourish, but the availability of natural sand has become extremely limited, owing in large part to the overexploitation of resources that has occurred in recent years, as has been the case in recent years.

Due to a scarcity of acceptable natural river sand, construction enterprises have been compelled to seek out suitable substitutes. Furthermore, because M-sand does not need to be carried across long distances, it has become a widely available product that has minimal impact on the environment.

It is possible to keep construction costs under better control.

Because of a scarcity of river sand, the price of this product has reached an all-time high in recent years. This means that the expenses of larger construction projects are not as simply predictable as they might otherwise be. Additionally, the expense of transportation has had a significant impact on the project's budgetary constraints.

the non-existence of silt and/or clay

Silt is the most serious danger to the durability of concrete, and it can be found in large quantities in natural river sand. The general percentage of silt allowed in such blends is less than 3 percent.

While most river sand contains between 5 and 20 percent silt, such high percentages result in concrete that is readily fragile and breaks when exposed to high temperatures. As a result, the amount of river sand that may be used for construction is minimal. Additionally, marine-derived materials such as shells and bark make up a portion of this sand's composition. Instead, M-sand does not include any of these contaminants, allowing constructors to have complete control over the quality of the concrete they are pouring into their structures.

Reduce the amount of by-product waste produced. Nothing is thrown away during the manufacturing process of M-sand. Indeed, lower-quality aggregates and minerals are treated and transformed into usable materials for use in the construction industry as well.

In the construction industry, the properties of M-Sand are important.
It is used to produce high-quality concrete.

In contrast to natural sand, manufactured sand must adhere to a strict set of norms and specifications in order to be used in construction. As a result, manufacturers may regulate the amount of contaminants present in the sand as well as its physical attributes. As a result, they are able to produce high-quality concentrate that is tightly controlled from beginning to end.

When used to increase the workability of concrete, it can be quite effective.
Quality of the finished product depends on the form, size, and texture of the sand that was used in the mixing of concrete. Manufacturers have control over these characteristics while making M-sand, and they can adjust them in order to boost the strength and longevity of the concrete they produce.

Defects are reduced when using high-quality sand.
When it comes to initial and ultimate setting times, M-sand is unsurpassed. Because the grains are precisely formed for the purpose of manufacturing high-quality concrete, the end product is less likely to contain common flaws and imperfections. Honeycombing, segregation, fracture, voids, capillarity, and bleeding are examples of these phenomena.

It is a more environmentally friendly option.
The demand for concrete increased in tandem with the expansion and development of the construction industry. As a result, construction companies have profited from the usage of natural sand, which is mined from river bed deposits. However, the excessive use of such a valuable resource has resulted in environmental calamities as well as river overexploitation.

Instead, because M-sand is being created in a lab, it does not necessitate the dredging of rivers or the destruction of natural regions in order to be produced. Additionally, the by-products of the manufacturing process can be employed in the building industry, reducing the likelihood of waste in the process.

It is cost-effective.
Natural sand is becoming increasingly scarce, which has resulted in an increase in the price of this precious resource over time. As a result, it has become increasingly difficult for construction companies to obtain high-quality items that are also within their budget.

Additionally, due to the fact that only a few rivers can still be mined for sand, the sand may need to be carried across considerable distances to reach the construction site on occasion.

Instead, M-sand is widely available and is within walking distance of the majority of construction sites. This makes it far easier to keep costs under control.

In the construction industry, M-Sand is preferred over river sand.
In contrast to M-sand, river sand contains contaminants that are primarily derived from the surrounding natural environment. Clay, silt, and marine goods are frequently found in this category. River sand is very difficult to dig, and only a small fraction of it is appropriate for use in the construction of roads and buildings.

Consequently, sand mining from river beds can be exceedingly destructive to the ecosystem, and once the results have been exploited, there will be no other use for the material.

M-sand, on the other hand, is a produced product that is the result of a meticulous design process. Although the material is still in its early stages and no long-term studies have been done, it has been evaluated for durability and strength. In this way, it can be considered not only excellent for construction, but also a preferred alternative to river sand in some situations.

Conclusion

M-sand is a sort of synthetic sand that has begun to take the place of natural river sand in several applications. This is primarily due to the global scarcity of this particular substance, which is typically more expensive and difficult to obtain. As an alternative, M-sand, which is produced by crushing a variety of rocks, quarry stones, and aggregates, is readily available, more environmentally friendly, and favourable for building enterprises.

Despite the fact that its coarse texture and the presence of micro-fine pchapters may significantly reduce its workability, this sand is used to produce concrete with excellent strength and durability.

www.ingramcontent.com/pod-product-compliance
Lightning Source LLC
Chambersburg PA
CBHW060826220526
45466CB00003B/999